MANAGING
CLIMATE
CHANGE

Papers from the
GREENHOUSE 2009 Conference

Imogen Jubb, Paul Holper and Wenju Cai

CSIRO
PUBLISHING

National Library of Australia Cataloguing-in-Publication entry

Managing climate change: papers from the GREENHOUSE 2009 Conference/editors, Imogen Jubb ;

Paul Holper ; Wenju Cai.

9780643098312 (hbk.)

Includes index.
Bibliography.

Global warming – Congresses.
Climatic changes – Congresses.
Global temperature changes – Congresses.
Climatic changes – Environmental aspects.
Climatic changes – Social aspects.

Jubb, Imogen.
Holper, Paul N.
Cai, Wenju.

363.73874

Published by
CSIRO PUBLISHING
150 Oxford Street (PO Box 1139)
Collingwood VIC 3066
Australia

Telephone: +61 3 9662 7666
Local call: 1300 788 000 (Australia only)
Fax: +61 3 9662 7555
Email: publishing.sales@csiro.au
Web site: www.publish.csiro.au

Set in 10/12 Adobe Minion and ITC Stone Sans
Edited by Janet Walker
Cover by The Modern Art Production Group
Text design by James Kelly
Typeset by Desktop Concepts Pty Ltd, Melbourne
Index by Russell Brooks
Printed in China by 1010 Printing International Ltd

The paper this book is printed on is certified by the Forest Stewardship Council (FSC) © 1996 FSC A.C. The FSC promotes environmentally responsible, socially beneficial and economically viable management of the world's forests.

CSIRO PUBLISHING publishes and distributes scientific, technical and health science books, magazines and journals from Australia to a worldwide audience and conducts these activities autonomously from the research activities of the Commonwealth Scientific and Industrial Research Organisation (CSIRO).

The views expressed in this publication are those of the author(s) and do not necessarily represent those of, and should not be attributed to, the publisher or CSIRO.

Foreword

GREENHOUSE 2009, the fifth conference in the GREENHOUSE series, was a busy four days. More than 500 delegates heard over 150 verbal presentations, read more than 70 posters, and consumed the sobering updates from the international climate change conference held in Copenhagen just a fortnight earlier.

A wealth of information and opinion about climate change was presented and I won't pretend to summarise all of that information. Rather, I will note just five of the main threads that were woven throughout the conference and reflect on some of the changes we've seen around the science and perception of climate change since GREENHOUSE 2007 and the challenges we should tackle before GREENHOUSE 2011.

First, we heard repeatedly that many of the metrics by which we measure modern climate change are tracking at – or above – the worst-case scenarios considered possible just a couple of years ago. Atmospheric greenhouse gas concentrations and sea level rise, for example, exceed what was predicted for them this decade. These metrics and others are projected to increase more rapidly than anticipated in the IPCC Fourth Assessment Report, released just two years ago. It was reported from Copenhagen that most empirical signals now indicate that trends in our climate, over and above the year-to-year fluctuations, are close to – or above – the worst-case scenarios predicted previously in the IPCC reports. We are confident about these statements because of efforts over recent years to build an expanding array of observations of our environment, although it was noted that there remain gaps in our knowledge in need of urgent clarification.

It is perhaps this unequivocal recognition of the magnitude of recent climate change that leads to my second thread.

We have witnessed a significant shift in the discourse around climate change. Just a couple of years ago, the public debate was focused on arguments about whether climate change was real, whether humanity contributed to it, and whether it was important. We now find ourselves discussing what we are going to do about climate change, through mitigation and adaptation. We have seen governments in Australia and abroad fund research and development into climate change mitigation strategies and adaptation measures. This funding shows recognition of the importance of the climate change phenomenon. Indeed, our own Minister for Climate Change, the Honorable Senator Penny Wong, when opening the GREENHOUSE 2009 conference, announced a major research effort to assist our Pacific neighbours respond to climate change – the $20 million Pacific Climate Change Science Program.

Most of the international attention to climate change now largely takes as read that climate change is a problem to which humanity is a major contributor. It is generally accepted that climate change is, or soon will be, a major problem for communities worldwide. Unlike previous changes in climate, modern climate change is occurring in the context of a vast human population living basically sedentary lives and relying on agriculture and built infrastructure for survival. For example, around 150 million people live in permanent settlements within one metre of current sea level. It is sobering to consider what the most recent projections of sea level rise, by as much as 1.5 metres this century, will mean for those people.

Recognition of the magnitude of the climate change challenge is an important shift, but we need to be careful that sight is not lost of the questions that remain to be answered. Many of these questions are around what climate change will mean for communities, industries and governments locally and in the next few decades. Ongoing research is required to characterise these local, lifetime effects better. It is difficult for us to design effective adaptation or mitigation strategies if we have only poor understanding of the specific changes we face, and when and where we will face them.

We heard throughout the conference that Australia is in many ways a climate change hotspot, but we also heard that many in the community remain unclear about what that will mean for them personally. Climate change remains a somewhat mysterious and uncertain phenomenon for many people despite the copious media coverage over recent years.

The third and fourth threads I have taken from the conference, around research and communication, are about improving our understanding of climate change impacts at local and decadal scales, within science and across the community.

The third thread reflects the messages from many speakers that we still have a lot to learn about our climate. We still strive to unravel fully the key climate processes that will change as we continue to load the atmosphere with greenhouse gases. We need to strengthen our research efforts to understand key climate fundamentals, such as the El Niño–Southern Oscillation (ENSO) and the Indian Ocean Dipole (IOD) that are so important to Australia, to improve projections of future climate dynamics. There remain significant challenges in downscaling climate projections to scales in space and time where people most want advice from science. Several speakers noted that we cannot expect to completely remove the uncertainties in climate projections but we should strive to better resolve some of those uncertainties that will most affect our forecasts of future climate. This will require enhanced efforts in both observational and climate modelling research and ongoing investment in research infrastructure. Several speakers reminded us of the need to continue to develop our nascent research into viable social and economic responses to climate changes that are now accepted as inevitable over coming decades.

Uncertainty is central also to the fourth thread I've taken from GREENHOUSE 2009. Communication. I heard many people speak of the difficulty of communicating the meaning and implications of climate change to the community. Many in the community have only the vaguest appreciation of what climate change is all about or why – or indeed whether – it should matter to them. There was mention of the disconnect between the language of science around climate change and the messages that people in the community, in industry and in government seek in order to understand and respond to climate change. As a science community, we continue to tell people about changes in global average temperature or sea level rise over the next 100 years, but those metrics don't connect with the information people need to make decisions locally over the next one to two decades. 'Global' and 'end of this century' are simply too distant for most people to identify with or care about.

South African speaker, Guy Midgely, also noted that it is easy to be overwhelmed by the sense of inevitability and crisis conveyed by much of the climate change discourse. Part of the problem is that our communication addresses the global nature of climate change, with insufficient attention to the local consequences or actions that can be taken by local communities. The 'think globally, act locally' message is rather out of balance.

A challenge for us all is to learn from the people faced with adaptation and understand what information they need to make decisions. We then need to do a better job of articulating climate change effects in the terms relevant to those people and to their needs. For example, many owners of coastal infrastructure aren't as interested in what mean sea level will be in 2100 as they are in how often key facilities will be inundated by the sea over the next few decades. These inundations are products of extreme events – not just the incremental elevation in average sea level. How often extreme events, such as extremes of local sea level, are likely to recur drives many of our planning and building guidelines in coastal communities.

Frequencies of extreme events will change dramatically as average sea level rises, but we have failed to explain to people how and why this will happen or what it will mean for existing planning and building guidelines. We need to do a better job of articulating how rising average sea level affects the extreme events that will threaten facilities with increasingly frequent inundation.

Finally, several of our plenary speakers reminded us that climate change is a particularly difficult policy, social and economic problem. It is a problem characterised by uncertainty, complexity, urgency and inequity. It is a problem outside our prior experience and for which we don't have established mechanisms to respond. It is also a problem that has the very real potential to be fundamentally disruptive to how we live and work and, for some, affect whether their societies survive.

Our responses to climate change inevitably will be affected by a multitude of other political, economic, social and psychological drivers. Many of those other drivers will be conceptually or emotionally more familiar to us than climate change. They will be easier or more important to deal with than climate change. They will apparently demand more immediate attention. For example, at the time of GREENHOUSE 2009, most people were probably far more concerned about the global economic crisis than they were about climate change. When the economic crisis passes, there inevitably will be some other urgent attention grabber that is easier for us to identify with than climate change.

A key challenge for us all is to bring local and immediate relevance to the discussion of climate change. We need to find ways to put climate change in the same frame as the many other issues that drive our political, economic, social and individual decision-making. Responses to climate change cannot be advanced in isolation from the other challenges facing societies around the world, but they almost certainly will be at least as important to our quality of life in coming decades.

It was also noted that we will need a mix of top–down, policy-driven actions and bottom–up, community actions aided by significant technology innovations. We heard throughout the week about some significant progress in some of these areas, including in the way we design and build the infrastructure on which we are so dependent and, of course, in the carbon pollution reduction legislation then being drafted. We heard that there *is* a growing sense of urgency about the need for action on climate change. Many speakers noted, however, that we have a way to go yet to harness that energy effectively for action across society and avoid the prospect of being crippled by complacency, despair or resignation.

Conferences such as GREENHOUSE 2009 won't solve the challenges of climate change. They do provide, however, useful periodic points of reference and reflection on whether we are on the right path to effectively tackle those challenges.

Bruce Mapstone
Chief of CSIRO Marine and Atmospheric Research

Contents

Foreword   *iii*

Preface   *xi*

List of contributors   *xv*

1. **Climate change: are we up to the challenge?**   1
    Graeme Pearman and Charmine Härtel

2. **Climate change and the Great Crash of 2008**   17
    Ross Garnaut

Part 1   Climate change science   **29**

3. **Twenty years of Australian Climate Change Science Program research**   31
    Paul Holper

4. **Tropical Australia and the Australian monsoon: general assessment
    and projected changes**   39
    Aurel Moise and Robert Colman

5. **Recent and projected rainfall trends in south-west Australia and the
    associated shifts in weather systems**   53
    Pandora Hope and Catherine Ganter

6. **How human-induced aerosols influence the ocean–atmosphere
    circulation: a review**   65
    Tim Cowan and Wenju Cai

7. **Freshwater biodiversity and climate change**   73
    Jenny Davis, Sam Lake and Ross Thompson

8. **Causes of changing southern hemisphere weather systems**   85
    Jorgen Frederiksen, Carsten Frederiksen, Stacey Osbrough and Janice Sisson

Part 2   Impacts and adaptation   **99**

9. **Australian agriculture in a climate of change**   101
    Mark Howden, Steven Crimp and Rohan Nelson

10. Wheat, wine and pie charts: advantages and limits to using current variability to think about future change in South Australia's climate   113
Peter Hayman and Bronya Alexander

11. Managing extreme heat in the vineyard: some lessons from the 2009 summer heatwave   123
Leanne Webb, John Whiting, Andrea Watt, Tom Hill, Fiona Wigg, Greg Dunn, Sonja Needs and Snow Barlow

12. Getting on target: energy and water efficiency in Western Australia's housing   137
Carolyn Hofmeester and Brad Pettitt

13. Sustainable energy as the primary tool to ameliorate climate change   145
Ray Wills

14. A national energy efficiency program for low-income households: responding equitably to climate change   155
Damian Sullivan and Josie Lee

15. Applying a climate change adaptation decision framework for the Adelaide–Mt Lofty Ranges   167
Douglas Bardsley and Susan Sweeney

16. Responding to oil vulnerability and climate change in our cities   177
Peter Newman

17. Managing climate risk in human settlements   185
Benjamin Preston and Robert Kay

18. Adapting infrastructure for climate change impacts   197
Michael Nolan

19. A critical look at the state of climate adaptation planning   205
Benjamin Preston and Richard Westaway

Part 3   Communicating climate change   221

20. Rising above hot air: a method for exploring attitudes towards zero-carbon lifestyles   223
Stefan Kaufman

21. Investigating the effectiveness of Energymark: changing public perceptions and behaviours using a longitudinal kitchen table approach   237
Anne-Maree Dowd and Peta Ashworth

22. **Talking climate change with the bush** 249
Clare Mullen, Shoni Maguire, Neil Plummer, David Jones and Colin Creighton

23. **Using Google Earth to visualise climate change scenarios in south-west Victoria** 259
Christopher Pettit, Jean-Philippe Aurambout, Falak Sheth, Victor Sposito, Garry O'Leary and Richard Eckard

Index *273*

Preface

A number of international, high-level science and policy meetings have been influential in the ongoing global climate change negotiations.

One of these meetings was GREENHOUSE 2009, where those involved in research, policy and communication of various aspects of climate change provided the latest assessments of the science and likely impacts on Australia and the world.

The first GREENHOUSE conference in the series was held in 1987. CSIRO's Dr Graeme Pearman played a major role in initiating that meeting. The aim was to promote public awareness, education, communication and understanding of the enhanced greenhouse effect by Australian decision makers. Since then, a number of GREENHOUSE conferences have taken place: GREENHOUSE 1994 in Wellington, New Zealand; GREENHOUSE 2005 in Melbourne; GREENHOUSE 2007 in Sydney; and GREENHOUSE 2009 in Perth.

Twenty-two years after that first meeting, Dr Pearman has provided one of the keynote papers of this publication – describing the important roles that behavioural and social sciences will have in determining responses to climate change.

Professor Ross Garnaut writes on the balance required between economic prosperity and accompanying reductions in global greenhouse gas emissions to prevent shifting society beyond its established patterns and coping range.

Further papers provide a summary of the state of climate change science, approaches to handling the effects and adaptation measures we are likely to face, and how to communicate the issue in order to generate better decision making and behavioural change towards sustainability.

This book provides an important snapshot of the concepts and ideas presented at the GREENHOUSE 2009 conference. PowerPoint presentations from the conference can be accessed online at <www.greenhouse2009.com>. Our thanks go to all those who provided such valuable contributions to the conference, which was attended by over 500 delegates.

The editors are very grateful to the authors for their contributions. All papers presented here were reviewed by at least two referees before acceptance. A list of referees appears on page xiii.

GREENHOUSE 2009 could not have occurred without the support and assistance from sponsors:

Australian Government Department of Climate Change
Government of Western Australia
Land and Water Australia
New Scientist
Maunsell AECOM
CSIRO
Asia–Pacific Network for Global Change Research
Australian Government Department of Agriculture, Fisheries and Forestry

The announcements, ideas and discussions at GREENHOUSE 2009 continue to make an important contribution to addressing and tackling climate change. We are confident that you will find the selection of papers presented in this book informative and revealing.

Imogen Jubb, Paul Holper and Wenju Cai

Program Committee

Dr Bryson Bates, Centre for Australian Weather and Climate Research
Dr David Bowran, WA Department of Agriculture
Mr Paul Holper, CSIRO
Ms Mandy Hopkins, CSIRO
Mr Sean Lucy, PWC
Ms Jo Mummery, Department of Climate Change
Dr Scott Power, Centre for Australian Weather and Climate Research
Mr Steve Waller, WA Department of Environment and Conservation

Organising Committee

Mr Tim Cowan, Centre for Australian Weather and Climate Research
Ms Edwina Hollander, CSIRO
Mr Paul Holper, CSIRO
Ms Mandy Hopkins, CSIRO
Ms Imogen Jubb, Australian Climate Change Science Program
Dr Zoe Loh, Centre for Australian Weather and Climate Research
Mr Ian Macadam, CSIRO
Dr Aurel Moise, Centre for Australian Weather and Climate Research

The editors gratefully acknowledge the assistance of the following external reviewers.

Andrea Bunting	RMIT University
Benjamin Preston	CSIRO
Greg Foliente	CSIRO
Karen Pearce	Bloom Communication
Leanne Webb	CSIRO
Linden Ashcroft	Australian National University
Luis Rodreguiz	CSIRO
Matthew Levinson	CSIRO
Paul Graham	CSIRO
Peter Christoff	University of Melbourne
Samuel Wilson	Monash University
Stefan Kaufman	Environmental Protection Authority, Victoria

List of contributors

Bronya Alexander
South Australian Research and Development Institute

Peta Ashworth
CSIRO Energy Transformed National Research Flagship

Jean-Philippe Aurambout
Future Farming Systems Research, Department of Primary Industries, Victoria

Douglas Bardsley
Geographical and Environmental Studies, University of Adelaide

Snow Barlow
School of Agriculture and Food Systems, University of Melbourne

Wenju Cai
CSIRO Wealth from Oceans National Research Flagship

Robert Colman
Centre for Australian Weather and Climate Research, a research partnership between the Bureau of Meteorology and CSIRO

Tim Cowan
CSIRO Wealth from Oceans National Research Flagship

Colin Creighton
Managing Climate Variability, a partnership across the Rural Research & Development Corporations

Steven Crimp
CSIRO Climate Adaptation National Research Flagship

Jenny Davis
Australian Centre for Biodiversity and School of Biological Sciences, Monash University

Anne-Maree Dowd
CSIRO Energy Transformed National Research Flagship

Greg Dunn
School of Agriculture and Food Systems, University of Melbourne

Richard Eckard
Future Farming Systems Research, Department of Primary Industries, Victoria

Carsten Frederiksen
Centre for Australian Weather and Climate Research, a research partnership between CSIRO and the Bureau of Meteorology

Jorgen Frederiksen
Centre for Australian Weather and Climate Research, a research partnership between CSIRO and the Bureau of Meteorology

Catherine Ganter
Centre for Australian Weather and Climate Research, a research partnership between CSIRO and the Bureau of Meteorology

Ross Garnaut
University of Melbourne

Charmine Härtel
Department of Management, Monash University

Peter Hayman
South Australian Research and Development Institute

Tom Hill
School of Agriculture and Food Systems, University of Melbourne

Carolyn Hofmeester
Murdoch University

Paul Holper
Australian Climate Change Science Program, CSIRO Marine and Atmospheric Research

Pandora Hope
Centre for Australian Weather and Climate Research, a research partnership between CSIRO and the Bureau of Meteorology

Mark Howden
CSIRO Climate Adaptation National Research Flagship

David Jones
National Climate Centre, Bureau of Meteorology

Stefan Kaufman
Environmental Protection Agency, Victoria

Robert Kay
Coastal Zone Management Pty Ltd

Sam Lake
Australian Centre for Biodiversity and School of Biological Sciences, Monash University

Josie Lee
Brotherhood of St Laurence

Shoni Maguire
National Climate Centre, Bureau of Meteorology

Aurel Moise
Centre for Australian Weather and Climate Research, a research partnership between CSIRO and the Bureau of Meteorology

Clare Mullen
National Climate Centre, Bureau of Meteorology

Sonja Needs
School of Agriculture and Food Systems, University of Melbourne

Rohan Nelson
Department of Climate Change, Canberra

Peter Newman
Curtin University Sustainability Policy Institute

Michael Nolan
AECOM Australia

Garry O'Leary
Future Farming Systems Research, Department of Primary Industries, Victoria

Stacey Osbrough
Centre for Australian Weather and Climate Research, a research partnership between CSIRO and the Bureau of Meteorology

Graeme Pearman
Graeme Pearman Consulting Pty Ltd; Monash University: Department of Geography and Environmental Science, Faculty of Business Economics and Monash Sustainability Institute

Christopher Pettit
Future Farming Systems Research, Department of Primary Industries, Victoria

Bradley Pettitt
School of Sustainability, Murdoch University; Mayor, City of Fremantle

Neil Plummer
National Climate Centre, Bureau of Meteorology

Benjamin Preston
CSIRO Climate Adaptation National Research Flagship

Falak Sheth
Future Farming Systems Research, Department of Primary Industries, Victoria

Janice Sisson
Centre for Australian Weather and Climate Research, a research partnership between CSIRO and the Bureau of Meteorology

Victor Sposito
Future Farming Systems Research, Department of Primary Industries, Victoria

Damian Sullivan
Brotherhood of St Laurence

Susan Sweeney
Department of Water, Land and Biodiversity Conservation, South Australia

Ross Thompson
Australian Centre for Biodiversity and School of Biological Sciences, Monash University

Andrea Watt
School of Agriculture and Food Systems, University of Melbourne

Leanne Webb
School of Agriculture and Food Systems, University of Melbourne; Centre for Australian Weather and Climate Research, a research partnership between CSIRO and the Bureau of Meteorology

Richard Westaway
CSIRO Climate Adaptation National Research Flagship

John Whiting
John Whiting Viticulture Consulting

Fiona Wigg
School of Agriculture and Food Systems, University of Melbourne

Ray Wills
Western Australian Sustainable Energy Association; School of Earth and Environment, University of Western Australia

1. CLIMATE CHANGE: ARE WE UP TO THE CHALLENGE?

Graeme Pearman and Charmine Härtel

Abstract

Climate change has been described as a diabolical issue because much of what we would like to know in order to manage the risks it presents is uncertain, the issues are complex and, without action, there is a potential for dangerous consequences. Policy development considers what the impact of greenhouse gas emissions might be, what response options are acceptable and what options are equitable. Together, these questions pose substantial challenges to individuals, companies and governments.

These challenges are, however, exacerbated by the lack of understanding in the general community of the different goals of science and risk management, and by the nature of human behaviour and societal institutions and norms. With respect to the first issue, we underscore in this chapter that 'truth' is the pursuit of the scientific approach, and thus definitive conclusions from the scientific community are typically *lagging indicators* aimed at providing certain explanations of what caused something to occur. In contrast, the risk management approach adopts a pragmatic and proactive stance (*leading* approach) aimed at sensibly balancing the probabilities of an event occurring against the impact of that event should it occur. This disparity of purpose, we argue, often leads the general community to underestimate the practical risks implicated by scientific conclusions. Adding to this barrier to action on the climate change issue is the nature of human behaviour and societal institutions and norms. We elucidate this point by providing a summary of the components of physical science, technology and economic research that have contributed to current understanding of the issue, and mapping on to it the components of behavioural and social science that appear most likely to be relevant to our understanding.

Our analysis concludes that the nature of the climate change issue requires a holistic capturing of current knowledge within the sub-disciplines of these fields and, in some cases, the need for new research activities. In the end, the most important lesson learned from the climate change issue may be that societal evolution has led our behaviour and societal institutions in directions that are unsustainable not only through changing climate, but in many other ways.

Introduction

Garnaut (2008) describes climate change as a 'diabolical' issue because it is uncertain in its format and extent, insidious rather than (as yet) confrontational, long-term rather than immediate, international as well as national and, in the absence of effective mitigation, carries a risk of dangerous consequences. Other dimensions to the diabolical nature of the issue include its sheer complexity, the urgent need for action and the inequities of its causes and effects.

In addressing the issue, governments and corporations respond by considering three questions: what is possible, what is acceptable, and what is equitable? (Pearman 2008; Härtel and Pearman 2010). With regards to what is possible, what are the geophysical realities of the cycling of greenhouse gases, the response of the climate system to these gases and the human inputs to emissions, and the direct impacts of climate changes on natural and human systems? Second, in terms of what is socially acceptable, how might the risks be managed, and what are the opportunities and potential for adaptive responses? Finally, in terms of what is equitable, how do we share the responsibility for action, establish agreements and legislation to share costs and manage inequities nationally and internationally?

This delineation of the major questions provides a convenient way of describing what is otherwise a complex of interactions without discernable structure. What is striking is the degree to which human behaviour and human governance systems form an integral part of how we contribute to, perceive and respond to the climate change issue. As observed by Shove (2006) in relation to the issue of energy consumption and climate change, response necessitates that we go beyond physical science and consider the values, norms and institutions that drive behaviour.

This requirement extends far beyond just energy consumption and incorporates how we perceive the problem, how we manage it, weigh options, and share responsibilities. Perhaps more relevant is the relatively poor state of our incorporation, until recently, of how human factors have influenced our response to the issue of climate change. It is argued in this chapter that without greater focus on these factors, there is a serious chance that we will not be up to the challenge of timely response to the potential dangers of climate change.

Vlek and Steg (2007) recently argued that environmental problems are basically social and behavioural problems. In response to the question, 'Are there social limits to adaptation to climate change?', Adger *et al.* (2009) suggested that the limits to adaptation are endogenous to society and hence contingent on ethics, knowledge, attitudes to risk and culture. Similarly, Hulme (2009), in addressing the question, 'Why do we disagree about climate change?', identified a range of behavioural and societal factors, such as perhaps science is not doing the job we expect of it; that we each have different values and views concerning our duty to others, to nature and to our deities; that we understand development differently, and that we seek to govern in different ways.

Härtel and Pearman (2010) recently examined the role of behavioural science in the climate change issue in terms of what is possible, acceptable and equitable. They first identified a wide range of physical science issues that relate to the climate change, and then mapped on to this the relevant fields of behavioural and social science. This chapter draws heavily on the work reported in Härtel and Pearman (2010) to explore who we are, how we behave and how our societal constructs interface with the issue.

Behavioural and societal constraints

Behavioural and social characteristics that may have influenced the development and management of the climate change issue may be considered at a number of levels. First, there are the factors that influence the behaviour of individuals deriving variously from genetic predispositions through to personal experiences and cultural norms. Second, there are the social structures that have evolved over time that organise and control communities, such as social institutions and political processes. The third characteristic is the interface between the first two, the individual responses reflecting both personal motivation and beliefs and social norms. It is beyond the scope of this chapter to review all of the ramifications of these levels in terms of the climate change issue, but we seek to highlight some of these areas.

Much of what is written reflects the first author's experience in the physical science of the climate change issue and at times, may show a lack of knowledge of the relevant psychological and behavioural literature. At the same time, however, it is hoped that the coupling of these physical science views with those of the behavioural science community may bring about new opportunities for research and understanding of how individual and communal behaviour has shaped our response to the issue thus far and opportunities for climate change-mitigation responses.

What is possible?

Examining what is possible for future climate change has been the objective of the physical science community for several decades. Developing this understanding involves empirical and theoretical knowledge of the global climate system and the examination of the response of physical (e.g. ice sheets, hydrology and sea level), biological (e.g. ecosystems and individual species) and community systems (e.g. water supplies, agriculture, human health) that are dependent on climate to understand how each may respond to future climate change. Some of the behavioural and social issues related to this process include the transfer of knowledge (advice), the nature of scepticism, innate and acquired behavioural tendencies, and attitudes and implicit beliefs.

Knowledge transfer

Physical science has identified probable changes to the climate system and how this may impact on matters of societal importance (IPCC 2007). This does not, however, guarantee that such information will be transferred to the wider community and thus influence governance and behavioural change. The knowledge transfer process (which applies more generally to the transfer of any expert advice) is complex. It is largely *ad hoc*, whereby many and diverse mechanisms are used, and for which there are no rules of engagement. The transfer itself is thus open to manipulation and misuse. The way science is conducted, particularly within the framework of the so-called purchaser–provider model,[1] means that there can be an unhealthy connection between anticipated and assigned outcomes of research, and the freedom and independence for innovation of the research itself. The situation relates to current practices and accepted norms of the role of accountability and economic outcomes from research investment.

The Intergovernmental Panel on Climate Change (IPCC) was developed as a mechanism for transferring complex, technical and multidisciplinary information from the science community to public and private sector decision makers. It was deliberately formulated to cross the boundaries of pure science and science application. This has attracted compliments from those who understand just how difficult this process is, but also derision from others who confuse the goals and roles with those of the scientific peer-review journal publication process. It is noteworthy that some of the most vocal critics of the IPCC assessments are themselves minor contributors to either process.

The role of the media is a special case. Journalistic adherence to the right of equal time (no matter how plausible, probable or informed one view may be) is juxtaposed with the commercial desire for conflict in reporting as a means of generating interest. Knowledge generated from research eventually enters the wider community understanding through education processes of several kinds: informal public education (media), formal and workforce. The

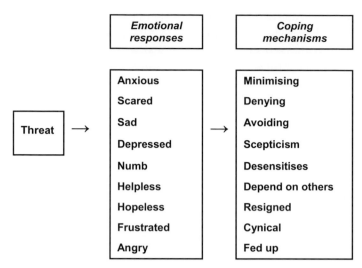

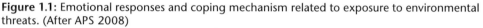

Figure 1.1: Emotional responses and coping mechanism related to exposure to environmental threats. (After APS 2008)

construction of curricula and the insurance of rigour in this transmission are important and different in each form and for each level of education, community, primary, secondary, tertiary or workplace. Aside from the inevitable preconceived biases of the educators, the processes themselves potentially contain significant delays that may influence the timeliness of the delivery of information and therefore the potential for informed public debate and democratic involvement in a rapidly moving issue.

Nature of scepticism

Scepticism is the cornerstone of good science. Scepticism is one of many coping responses to an equally diverse set of emotional responses to the threats of environmental change (Figure 1.1, APS 2008). Often experiential evidence – one's personal exposure to and interpretation of the world – will dominate an individual's beliefs and world view, irrespective of scientific or other expert advice to the contrary. Scepticism can be, on the one hand, a crutch to avoid a clash with personal views or to avoid the need to change a view or respond to the implications of a different view. On the other hand, scepticism may be a tool for confronting a world view that may be legitimately opposed by one's own sense of reality. In all cases, these are psychological characteristics that must be recognised as having the potential to affect the integration of expert advice into mental representations of the nature and consequences of climate change.

Innate and acquired behavioural tendencies

Each of us has different propensities for a variety of behaviours. These affect our level of acceptance of change, reflecting a range of behaviour predispositions including being conservative or radical, positive or negative, short term or strategic, etc. Hulme (2009), for example, describes individual tendencies towards fatalism, individualism, egalitarianism or hierarchy – reflecting ideas identified originally in the work of Douglas and Wildavsky (1982). These differences are not assessed as being of greater or lesser value, but rather simply, reflecting the fact that we are all different. Irrespective of whether these tendencies are innate or acquired through socialisation, they influence how we respond to environmental threats such as that of climate change.

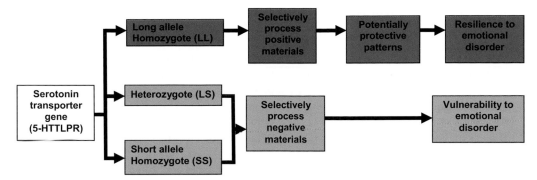

Figure 1.2: Identification of a single gene control over personal preference for positive or negative materials and the potential for this to influence behavioural resilience. (Based on Fox *et al.* 2009)

An example of where underpinning genetic influences appear to be important has being articulated by Fox *et al.* (2009) who found that the propensity of individuals to choose between positive or negative materials (photographic images from the International Affective Picture Set) was associated with particular alleles related to the serotonin transporter gene (Figure 1.2). It is conceivable that such embedded genetic variability generates a diversity of behavioural responses that have survival value, although this is not known.

Other underpinning drivers of behaviour will not be genetically based but result from experience gained through family, wider societal norms and education. Indeed it is most likely that many individual behavioural characteristics result from the interaction of both forms of influence. Cases where learning influences dominate include the formulation of beliefs and values derived from religious or cultural circumstances.

Attitudes and implicit beliefs

Attitudes and implicit beliefs are among of a number of unconscious antecedents of behaviour. These attitudes include the perception of success, beliefs and values, religious and cultural inheritance and, sometimes, the development of an ethos of sustainability. Such attitudes may be strongly entrenched, having been built over many years and influenced by family and peers, by education and wider information gathering. Existing attitudes may be consistent or inconsistent with those required in order to respond to the threat of climate change. For example, the perception of success, in current times, is often built around the acquisition of objects or material wealth – what Hamilton and Denniss (2005) describe as 'stuff'. Such attitudes may be based on notions of success and well-being and encouraged under the influence of advertising campaigns designed to sell objects, irrespective of their applicability to a changed world, or their contribution to that changing world, or including whether or not, and to what extent, they contribute to greenhouse gas emissions. A clear conflict exists between such attitudes and the intellectually accepted view that an individual's emissions will need to change to avoid further global warming.

In addition to individual attitudes, large numbers of individuals can share common world views such as those that dominate attitudes to population growth, consumerism, the role of markets, the role of financial institutions, etc. These underpin a largely unconscious set of belief structures and can potentially drive particular collective directions in community development and capacity for change. Saul (1997) argues that it is these 'ideological' views of the world, of which we are largely unconscious, that drive our civilisation. For example, in the case of the climate change issue, it is clear that consumerism (influenced by advertising) and the

Table 1.2   Relationships between areas of physical science, economics and energy technology that are involved in the determination of 'what is possible' in the context of climate change, and the kinds of behavioural and social science research areas and issues that are also relevant (based on Härtel and Pearman 2010)

Physical aspects of the climate system

Physical sciences		Areas of behavioural science	Specific behavioural issues
Biogeochemistry	Greenhouse gas levels	**Capturing** complex, uncertain, **advice** for policy development and community understanding	Complex, largely **ad hoc processes**; exposure to weaknesses and manipulation
			Role of science in modern societies, management and funding models
Physics and fluid dynamics of the atmosphere and oceans	**Climate change and variability**	**Basis of scepticism**, its role in science and in the wider community	Role of the **media** in transference or distortion of knowledge
	Observed changes to the physical climate system	**Factual versus emotive drivers** of acceptance or otherwise of expert advice	Role of **education** in public awareness & policy development
			Commitment to **experiential evidence** of reality versus descriptions based on rationale argument; what is **rational**
Dynamic response of the Earth's biology	**Complexity of ecosystems,** limited predictability and the changing role of conservation	**Emotive connections** to land and biodiversity	Role of **vested interests** in stifling or promoting change of wider community value
		Learned and inherited propensities	Drivers and constraints on **community attitudes towards threats**
		Dealing with **probabilities and assessment of risk**	**Holistic thinking** in the determination of options
			Acceptance of change
			Religious demography of **beliefs and attitudes**

Physical sciences	Areas of behavioural science	Specific behavioural issues
Future energy demand related to population growth, affluence and technological development **Population growth** and immigration policy **Available technologies** and current infrastructure investments	**Construction of attitudes** based on fact, manipulation and perception	Perceived **value of heritage**, environmental protection and international responsibility Role of **governments**, management of diabolical problems: uncertainty, complexity, time scales, equity Management of responsibilities across **competing economic sectors**, self-interest and jurisdictional regions Attitudes towards **population growth as a paradigm of success** and driver of emissions growth Attitudes to **wealth generation as paradigm of success**, consumer behaviour, materialism, post-materialism Attitudes related to **demography, wealth, religion, education**
Anticipate global climate change under alternative human societal futures **Economic costs** of new technologies; impact of carbon trading; scaling up	Growing **ethos of sustainability** and social/environmental responsibility of corporations, conservation	**Role of markets** in selection of strategic options and maintenance of resilience Role of government intervention in the **maintenance of resilience**

Economics and energy technology

change issue and available options for mitigation and adaptation will differ within and between specific societies, influenced by education, public communication and personal levels of empowerment that affect how individuals might be committed to, or encouraged to, responsive action.

Companies, governments and individuals each play different roles in causing and responding to climate change. Each assesses their responsibility in the context of alternative aspirations, roles and capacities, yet their combined actions are necessary for satisfactorily addressing the overall challenge.

Each economic sector, jurisdiction and socio-economic group is variously exposed to climate change and therefore will determine its respective exposures and risk differently including valuing of natural resources and biodiversity, level of altruism and humanitarian action. This means that responses will also need to be diverse in the treatment of components such as:

- education – formal, workforce and public;
- improved and new management practices (agriculture, water use, energy provisions, etc.);
- balance across all sectors and jurisdictions in the face of competition and open markets;
- changes to personal lifestyles and expectations, and
- protection in the transition from the status quo to a changed future state for exposed sectors and people.

Characteristics of an enlightened response

The climate change issue highlights a number of key characteristics that might be viewed as underpinning an enlightened response and these include holistic and strategic thinking, uncertainty, flexibility and resilience.

Holistic thinking

Climate change coexists with poverty alleviation, energy and national security, food, water, health, environmental security and disaster mitigation. Climate change cannot be considered solely an environmental issue. It affects biodiversity conservation, human-built infrastructure, capacity of wealth-generating processes such as agriculture and energy sourcing, the management of migration and of emergency services, and the security, health and well-being of societies. Solutions that mitigate or adapt to the effects climate change that are singular in their approach (that is, manage only costs or jobs or the environment), are likely to always deliver unwelcome results. The 21st century solution of problems demands holistic thinking to incorporate the multi-dimensional nature of most problems and the multidimensional outcomes that we aspire to achieve.

Strategic thinking

The climate change issue reinforces the view that the direction of human social evolution cannot be determined by human aspirations alone but will need to reflect the realities of the physical world. It is also true that human societies have the capability of envisaging a future world which delivers the wide range of aspirational needs largely shared across religious, national and other societal boundaries. These are aspirations related to human well-being, such as freedom, health, wealth, environmental sustainability and stewardship and spirituality. But if these aspirations are to be achieved, we need to build a shared clarity about where we wish to be over time, economically, socially and environmentally. This should not be set in stone but act as a guide towards broad and widely shared aspirations. The climate change issue invokes serious questions about societal evolution which has been devoid of direction and long-term strategy. Has it led our behaviour and societal institutions in directions that are unsustainable not only in terms of how our climate is changing but in many other ways? How might we incorporate into future societal evolution those constraints that will more probably ensure future resilience?

Uncertainty

The future will always be uncertain, as will the nature and effects of climate change. Given the complexity of how the physical climate system responds to increased greenhouse gases, and how that response delivers different impacts on natural and human systems in different regions, and how human systems in turn respond to those changes, uncertainty will always characterise this issue. The uncertainty derives from the complex cascade of interactions between each level of the problem and highlights the need to understand that action needs to be taken in the face of uncertainty. Rather than attract delay, uncertainty should demand urgent action, flexibility and precaution.

Flexibility

Uncertainties exist concerning the exact future of the global and regional climate, the options for adapting to change and the options for modifying our future energy systems. This demands that options are kept open. It is akin to investing in the share market. Rarely is there sufficient confidence that a particular stock will perform well over time, thus suggesting the use of a portfolio approach. We need to invest simultaneously in alternative ways of handling, for example, energy futures, so that through experience we will learn which of those options meet the criteria of responding to the climate change issue, provide for energy demands at competitive prices, satisfy other environmental conditions and respond to community perceptions of values and risks associated with each alternative. Flexibility provides options as new knowledge emerges and avoids dead-ends or undesirable dislocations. It may not be efficient in the short term, but it is the best way to cope with uncertainty and build resilience.

Resilience

Ultimately the objective in an uncertain world is to maintain resilience; the capacity under new conditions for individuals, societies, companies and ecosystems to continue to operate and not to succumb to the vagaries of those conditions. It is not that those operations will remain unchanged; on the contrary, that is very unlikely. But it is likely that they will change from their current state in a phased and continuous fashion towards an alternative state and structure in the future. Diversity and flexibility maximises the chance of dealing with unforeseen futures, minimising the chance of major discontinuities. It avoids systems approaches that can be so narrowly focused that they potentially lose the capacity to respond.

Conclusions

Physical scientists have largely dominated the development of our current understanding of the climate change issue. Some of us have suffered under the illusion that as knowledge of the physical world improves, this rationally based information will lead to rational responses that deal with the perceived threats. Indeed, it is possible that society in general suffers from this illusion, concluding that given a little more time and money, the answers will be provided and that this will solve our problems. In carrying out this brief mapping of the climate change challenges and the related human behavioural components (based on Härtel and Pearman 2010), it has become clear that there is no such guarantee. Behavioural responses are subject to individual and collective behavioural propensities, and these constrain the type, magnitude and rate at which it is likely changes can be made, even though these may appear to be clearly identified as necessary from a physical science perspective.

This leads to the conclusion that in addition to the number of issues that determine that the climate change issue is indeed a 'diabolical' problem, the psychological and behavioural components test our confidence that major impacts from climate change can be avoided. Climate change is a threat to the planet that has not previously existed and is of a scale that exceeds anything experienced previously. Its complexities, uncertainties and inequities, when added to these human dimensions of behavioural change, can be regarded as a threat that may not be resolved without significant long-term and serious impact. *We may not be up to the challenge.*

Many people who understand the risk of climate change are currently focusing on applied actions to modify human behaviour to reduce emissions and bring about adaptive preparedness. These activities need to be encouraged. However, there is a danger that in our enthusiasm, we may underestimate the diversity of the psychological and social melange into which we project advertising, promotional and regulation campaigns. This risks the possibility that they may fail to deliver timely responses for deep, often poorly understood, reasons of human behaviour. Much of this fundamental knowledge about these preconditions is available within the annals of behavioural and social science literature and needs to be more effectively captured, but in some cases there is a need for further knowledge development.

Building this information into a more holistic view of the climate change issue is urgent. Flexibility regarding options for action in the state of uncertainty is valuable as it provides the degree of resilience required for the community management of this global environmental change.

One might reasonably ask if had more of us, 20 years ago, asked such questions and demanded a reshaping of the research agenda with the inclusion of behavioural and social science in the development of the underpinning knowledge of climate change, how much better prepared we might we have been today?

Acknowledgements

The views expressed in this document are those of the authors and have been greatly influenced by their joint work (Härtel and Pearman 2010). That work was supported jointly by the Monash Sustainability Institute and the Faculty of Business and Economics. Also during this work, discussions between the first author and the following colleagues are acknowledged: Peta Ashworth, John Fien, Ralph Horne, Anna Littleboy, Janet Stanley, Jodi-Anne M Smith and Samuel Wilson. The first author is grateful for the way they shared their knowledge and were patient with a physical scientist exploring disciplines well beyond his personal expertise.

Endnote

1   The purchaser–provider model has been particularly developed within the health services sector and has not been without critics (see, for example, Street 1994), but it has also pervaded the wider science funding process in recent times, albeit not necessarily with apparent conscience intent.

References

Adger NW, Dessai S, Goulden M, Hulme M, Lorenzoni I, Nelson DR, Naess LO, Wolf J and Wreford A (2009) Are there social limits to adaptation to climate change? *Climatic Change* **93**, 335–354.

APS (2008) *Climate Change: What You Can Do.* Australian Psychological Society. <www.psychology.org.au/publications/tip_sheets/climate/>

De Kirby K, Morgan P, Nordhaus T, and Shellenberger M (2007) Irrationality wants to be your friend. In: *Ignition: What Can You do to Fight Global Warming and Spark a Movement.* (Eds JH Isham and SA Waage), Chapter 4. Island Press, Washington DC.

Douglas M and Wildavsky A (1982) *Risk and Culture: An Essay on the Selection of Technological and Environmental Dangers.* University of California Press, Berkeley, CA.

Fox E, Ridgewell A and Ashwin C (2009) Looking on the bright side: biased attention and the human serotonin transporter gene. *Proceedings of the Royal Society B* **276**, 1747–1751. doi:10.1098/rspb.2008.1788.

Garnaut R (2008) *The Garnaut Climate Change Report.* Cambridge University Press, Cambridge, UK.

Hamilton C and Denniss R (2005) *Affluenza: When too Much is Never Enough.* Allen and Unwin, Crows Nest.

Härtel C and Pearman GI (2010) The climate change issue: the role for behavioral sciences in understanding and responding. *Journal of Management and Organization* **16**(1), in press.

Hulme M (2009) *Why We Disagree about Climate Change: Understanding Controversy, Inaction and Opportunity.* Cambridge University Press, Cambridge.

IPCC (2007) Climate Change 2007: Synthesis Report. Contribution of Working Groups I, II and III to the Fourth Assessment Report of the Intergovernmental Panel on Climate Change. IPCC, Geneva, Switzerland.

Pearman GI (2008) Climate change: risk in Australia under alternative emissions futures. Document prepared as part of the study: *Climate Change Impacts and Risk: Modelling of the macroeconomic, sectoral and distributional implications of long-term greenhouse-gas emissions reduction in Australia* by the Australian Federal Treasury. <www.treasury.gov.au/lowpollutionfuture/consultants_report/downloads/Risk_in_Australia_under_alternative_emissions_futures.pdf>

Saul JR (1997) *The Unconscious Civilisation.* Penguin, Ringwood.

Shove E (2006) Efficiency and consumption: technology and practice. In: *The Earthscan Reader in Sustainable Consumption.* (Ed. T Jackson) Chapter 20. Earthscan, UK.

Street A (1994) 'Purchaser/provider and Managed Competition: Importing Chaos?' Working Paper 36, Centre of Health Program Evaluation, West Heidelberg, Victoria, Australia. <www.buseco.monash.edu.au/centres/che/pubs/wp36.pdf>

Vlek C and Steg L (2007) Human behaviour and environmental sustainability: problems, driving forces and research topics. *Journal of Social Issues* **63**, 1–19.

2. CLIMATE CHANGE AND THE GREAT CRASH OF 2008

Ross Garnaut

Abstract

Australia and the international community are living through a time of consequences.

According to mainstream science we have little time to stabilise and then reduce global emissions of greenhouse gases in order to provide a reasonable chance of avoiding dangerous climate change.

Mitigation too weak to avoid dangerous climate change could give human society a large shock to its established patterns of life, potentially moving society beyond its coping range. The decisions made at Copenhagen in 2009 will be fateful in deciding the dimension of change we face.

Good environmental outcomes will only be delivered if long-term economic prosperity is built on steady and large reductions in global greenhouse gas emissions. This chapter looks at five points of interaction between Australian climate change policy and the great financial crash of 2008. These points include the relationship between global recession and emissions growth; investment in structural change; financial fragility, and the political economy of employment change and the market economy. The role of Australian policy in contributing to strong global action is discussed in the light of these points of interaction.

Introduction

Australia and the international community are living through a time of consequences.

If the mainstream science is broadly right, we have little time to stabilise and then to reduce global emissions of greenhouse gases if there is to be a reasonable chance of avoiding danger-ous anthropogenic climate change. In 2009, a meeting under United Nations auspices is to set the rules for emissions reductions and trade for the period after the Kyoto arrangements come to their conclusion in 2012. This is the first year of a new Obama Administration in the United States, coming to office with a commitment to participation in strong global action to combat anthropogenic climate change. It is the year in which Australia makes critical decisions on mitigation policy.

It is also the year after the Great Crash of 2008. A collapse of global financial intermedia-tion of unprecedented dimension is sending powerful recessionary pulses through the world economy.

The concluding chapter of the Garnaut Climate Change Review stated that when human society receives a large shock to its established patterns of life, the outcome is unpredictable but generally problematic (Garnaut 2008, p. 591). Things fall apart.

Unmitigated climate change, or mitigation too weak to avoid dangerous climate change, could give human society such a shock.

We know that the possibilities from climate change include shocks far more severe than those in the past which have exceeded society's capacity to cope and moved societies to the point of fracture.

For these reasons, the decisions that are to be made at Copenhagen in December 2009 will be fateful - whether to press ahead with a comprehensive global agreement to mitigate climate change, or to postpone action, or to do nothing, or to take one more step on a journey of a thousand miles.

It happens that each of the three examples I gave in Chapter 24 of the Garnaut Climate Change Review of shocks that had knocked human society from its settled course were examples of financial instability. This reflected my longstanding professional interest in development and international finance, rather than the ominous news emanating from New York and London as I worked on my Review.

This financial shock of 2008 is one of the big events in history. As a financial shock, it may have been the biggest, although corrective policy action is moderating its continuing consequences for the real economy.

I presented the Final Report of the Climate Change Review to the Prime Minister on the morning of 30 September 2008. This was the morning after the night of the largest ever single day points fall on the New York Stock Exchange. The equities markets had been on edge as private credit shrivelled after the collapse of Lehman Brothers in mid-September. The Bush administration crafted a response. Market participants panicked when, on the United States business day of 29 September, our morning of 30 September, Congress rejected the President's recovery package.

After that, the panic gave way to heavy pessimism, which has begun to lift even while the developed countries remain in recession in mid-2009.

The Great Crash of 2008 and its recessionary aftermath have provided the whole context for Australian and international discussion of my Review. They have provided the whole context of discussion of the Government's December White Paper, and of the draft of the Government's Emissions Trading Scheme legislation.

Climate change was always going to be a diabolical policy problem (Garnaut 2008, p. xviii). I described it as being harder than any other issue of high importance that has come before our polity in living memory.

I wondered in the Final Report whether climate change policy was too hard for rational policy making. It was too complex. The special interests were too numerous, powerful and intense. The time frames within which effects become evident were too long, and the time frames within which action must be effected too short.

However difficult the climate change policy problem was in the first nine months of 2008 when I wrote those words, it is harder now.

It was a theme of my Review that there was no fundamental conflict between prosperity and the mitigation of climate change. It was an error to assert the priority of either economic or environmental values. Good economic policy and good environmental policy both required the removal of links between economic activity and greenhouse gas emissions.

There could be no victory for climate change mitigation that was not based on careful assessment of economic consequences. There would, on a balance of probabilities, be no long-term victory for economic prosperity that was not built on steady and large reductions in global greenhouse gas emissions, and eventually, their concentrations.

In the end, we would get both good economic and good environmental outcomes, or we would get neither.

This was true in high prosperity. It is true now, with the global economy in the aftermath of its largest proportionate decline in output since the 1930s, and of the largest proportionate decline ever in international trade over a similarly short period. It will be true in the challenging years that lie ahead.

Links between the Great Crash and climate change policy

Here I will discuss five links between the financial crisis and subsequent recession and climate change policy.

First, the decline in growth of economic activity reduces the rate of growth of emissions. This gives us a bit more time before dangerous points are reached.

Second, much capital and labour is underemployed in recession, and so they are available for new kinds of economic activity. This lowers the opportunity cost of investment in structural change. It is a time to invest in new technologies at relatively low cost.

Third, many enterprises are more fragile financially, and in a weak position to carry increased costs that they cannot pass on to users of their products.

These are probably the most important real economic interactions between climate change mitigation and the Great Crash.

There are two powerful political economy effects of the Great Crash and its aftermath, rendering more difficult the implementation of policy directed at strong reductions in greenhouse gas emissions at low cost.

First, there is greater anxiety about changes in the structure of employment of labour and capital. The inevitable losers from change are more vocal and effective in expression of that anxiety. The winners to a considerable extent do not yet know who they are, and so to that extent do not participate in the policy debate.

Second, recession weakens support for policies based on market exchange, the more so when its origins lie in the failure of markets. It therefore changes, for a while, the policies that are likely to be favoured in the wider polity, away from those that rely heavily on market exchange, towards those that rely on the exercise of discretion by governments.

Let us look more closely at these five points of interaction between the financial and economic crisis and the contemporary Australian discussion of climate change policy.

Global recession slows emissions growth

How large is the breathing space in growth in emissions? Unfortunately, small, especially compared with the scale and urgency of the mitigation task.

The Garnaut Climate Change Review drew attention to the powerful forces that were causing emissions growth in the early 21st century to be much more rapid than anticipated by the IPCC Reports. Economic growth was stronger; it was concentrated more in economies at stages of development in which growth was highly energy-intensive, and it was concentrated in economies, first of all China, India and Indonesia, in which coal was relatively abundant and cheap. Under 'business as usual', emissions growth through the 21st century was likely to be higher than in the so-called A1FI scenario – that with the highest rate of emissions growth. A1FI had previously been considered to be extreme, and the IPCC, and analysts such as Stern (2007) who drew on it, relied on scenarios with much lower rates of emissions growth.

How much is the emissions growth outlook changed by the Great Crash and its recessionary aftermath?

For the time being, emissions growth has slowed considerably. After a period in which average growth in the world economy, weighted by purchasing power, was near 5% per annum, output is set to fall in 2009. Growth in the major developing countries, while remaining positive, has slowed significantly.

In China, the world's largest and most rapidly growing major source of greenhouse gas emissions, the downturn came suddenly in the third quarter of 2008. China is now a market economy with its own business cycle. Exceptionally high investment in commercial real estate and urban housing was due for a correction by 2008. The correction coincided with a stunning decline in exports, as domestic demand collapsed in the United States, Europe, Japan and Korea. There was a fall for a while in production of such highly emissions-intensive products as electricity, metals and cement.

It will be some time before we have reliable data for global emissions in the period since the financial crisis began to bite deeply into real economic activity. Emissions probably fell a bit for the world as a whole in and after September 2008.

The trajectory of emissions growth in the years immediately ahead depends first of all on the timing and strength of recovery from recession. China is seeing a sharp increase in economic activity in response to large fiscal and immense monetary expansion. The effects are likely to be evident from the second quarter of 2009. The second and third most populous developing economies, India and Indonesia, are returning quickly to reasonably strong growth.

Elsewhere, recovery will be mostly slow, uneven, and in some cases delayed for a considerable time by continued weakness in financial intermediation.

With colleagues, I have said elsewhere that there may be no overall emissions growth for two or three years through the current recessionary episode (Garnaut *et al.* 2009; Garnaut 2010). The most likely course is for a return to growth that shifts back the curve of emissions levels over time by two or three years. This would mean that global emissions levels expected for 2030 would not be reached until 2032 or 2033 (Garnaut *et al.* 2008; Garnaut 2008).

The conditions for sustained strong global economic growth, led by China and the large developing countries, are likely still to be present after the financial and economic crisis.

There are several risks to the return to strong global growth, however, concentrated in the developed countries. The recession may leave a durable legacy of weakness in global financial intermediation, with continuing dampening effects on the scale and efficiency of trade and investment. It may leave a legacy of greater preference for inward-looking, protectionist approaches to economic development. More generally, it may leave a legacy of interventions in the economy that reduce productivity and incomes and the potential for their growth. It may leave a legacy of political instability in some countries and regions – although probably not China, India or Indonesia – which is inimical to growth. It may leave an ideological legacy of distrust of market exchange, which renders more difficult and less likely the adoption of productivity-raising reforms in many countries (Garnaut with Llewellyn-Smith 2009).

Any of these developments could weaken the beneficent processes of economic growth in the early 21st century, which have reduced the number of people living in poverty more rapidly than ever before in human history.

Any of these developments would lower rates of growth in energy use.

While any substantial diminution of future growth prospects may reduce 'business as usual' emissions from what they would otherwise have been, it would not be good news for reductions in emissions. It would make effective mitigation policies less likely. The overall impact would be to increase, not reduce, the risks of dangerous climate change.

Even in mid-2009, in deep global recession, with total emissions possibly lower than a year ago, the level of emissions is high enough for concentrations in the atmosphere to continue to rise at a considerable rate.

This is part of the context in which we should understand the 'breathing space' provided by the financial and economic crisis. The other part is the reality that concentrations of carbon dioxide equivalent in the atmosphere are already close to 450 parts per million (ppm).

Recession is a good time for investment in structural change

Global recession and the period of recovery that follows is a good time economically for investment in necessary structural change.

In global recession, there is too little effective demand to employ available resources in the world as a whole and in most of its parts. Good economic policy requires fiscal policy action to increase demand. Any increase in expenditure helps to reduce the immediate deficiency in demand. Expenditure that is focused on building an economy that will do well in future delivers an additional dividend: it augments future as well as current incomes.

It is for this reason that the fiscal responses to the current recession in many countries have included measures to reduce the emissions intensity of economic activity. China's stimulus packages include major commitments to renewable energy, public transport, and broadly to economisation on energy use. The Obama recovery package allocates large sums to support investment in low-emissions energy technologies and more efficient electricity transmission. European Governments have given high priority to encouragement of production and use of renewable energy. Australia's second recovery package gave high priority to economisation in energy use through insulation of housing. There is considerable discussion about increasing the climate change mitigation component of future stimulus packages in all of these countries, and others.

The case for investment in mitigation in recession is not that it will increase total employment more than alternative patterns of expenditure. Effective mitigation policy encourages the movement of resources from high-emissions towards low-emissions activities. Gradual structural change of the kind and dimension required for effective mitigation is unlikely to affect total employment one way or another. There is no reason to expect that, on balance, employment will be higher or lower as a result of this movement.

Gradual structural change expands employment in some areas and reduces it in others. Whether or not resources are fully employed depends on familiar macroeconomic considerations, some of which are strongly affected by policy.

The case for investment in mitigation in recession is that expenditure that improves the future operation of the economy gives value beyond the immediate stimulus to demand. Any public investment has lower opportunity cost than at other times. Investment in structural change towards a low-emissions economy, if designed and implemented well, is likely to have relatively high long-term value.

Be careful of financial fragility

The financial fragility of many enterprises is a reason to avoid sharp increases in costs until the economy has begun to expand after recession. Even if the whole of the increase in permit revenues received by the business sector were recouped from the passing on of cost increases or in other ways, there would be uneven distribution across firms.

Looking from March 2009, it would seem likely that the economy will be expanding again by the middle of 2011, when the Australian Government's proposed emissions trading scheme is scheduled to be introduced.

The political economy of rising and high unemployment

Whatever the overall effect of climate change mitigation policies on employment, there is much greater sensitivity about the loss of some jobs from policy changes at times of rising and high unemployment, and less confidence that change really will lead to commensurate growth in employment elsewhere in the economy. This generates doubts about policies that introduce pressures for contraction of some sources of employment, and expansion of others. It also generates fertile ground in which vested interests affected adversely by structural change can plant opposition to policy change.

Whatever the overall employment effects, there will be losers as well as winners from structural change. The losers are likely to be more vocal in expression of their concerns at any time, and more stridently so in recession. The beneficiaries of structural change are often silent – some for the good reason that they do not know who they are until after the new policies are in place.

So whether or not recession, and, more so, the recovery period that follows, is a good time for investment in structural change, as the basic economics suggests, it is a time when the political economy of reform is difficult.

There are historical cases of governments taking heed of the economic realities, and pressing ahead with reform despite the difficult political context created by recession. The outstanding Australian example is the Hawke Government's announcement in the depths of recession of the largest step in reduction of Australian protection in March 1991, with the new measures to take effect in what was expected to be, and was, a period of post-recession expansion.

It is more common for governments to give weight to the political difficulty than to the opportunity of reform in recession and its aftermath. Attempted reform in recession usually runs risks of compromise in response to business pressures which, unless designed with care, can reduce the value of the reform on return to prosperity.

The political economy of the market economy during and after recession

Most deep recessions leave some legacy of distrust of market exchange, and of increased sympathy for interventionist government policies. This recession is shaping up to be by far the deepest and longest since the 1930s. With its origins in failures at the heart of the global market economy in the New York and London banks, it would be surprising if it did not leave an unusually deep ideological legacy.

Some of the newly preferred interventions by government will be justified by lessons of experience. These will include the need for more effective regulation of transactions by deposit-taking banks.

Others will not. There will be greater resistance to reliance on markets in circumstances in which they contribute unambiguously to rising incomes. For reasons discussed in the Review, market-oriented approaches to mitigation are likely to secure larger reductions in emissions at lower costs than a myriad of interventions favouring some economic activities over others. But the setting for making the case for market-oriented approaches will be more difficult in recession and its aftermath.

It must be said that the reality as well as the perception of the value of market exchange may be diminished by deep recession, to the extent that it reduces the capacity, competitiveness or efficiency of the financial sector. If owners of capital, or intermediaries in the exchange of capital, have less capacity and willingness to take risks in holding new financial instruments, the forward market for permits will reveal a steeper contango and be less efficient than would otherwise be the case. This is a reason for ensuring that the financial sector has returned to health before relying on it to set the price for emissions permits.

Global recession and Copenhagen

The global recession provides a difficult political context for preparations for Copenhagen in December 2009.

Other developments and conditions are supportive of a strong outcome at Copenhagen. These include the change of Administration in the United States, the Australian Government's recent signing of the Kyoto Treaty, the understanding in Hong Kong, Korea and Taiwan that these economies will need to accept mitigation commitments in line with their contemporary economic status, and increasing evidence from the scientific community of the urgency of climate change mitigation. There is considerable momentum in domestic mitigation activity and discussion in many countries, including China, Indonesia and South Africa amongst major developing countries. Domestic opinion in most countries remains strongly supportive of mitigation, although its priority has fallen relative to the maintenance of employment and incomes.

An effective global agreement would have five parts.

First, it would embody an understanding on the desired level of ambition in global mitigation, expressed as a desired trajectory for reduction of emissions over time.

Second, it would allocate the emissions budget embodied in this trajectory across countries as emissions entitlements. This would be based on clear principles. The Review judged that convergence towards equal per capita entitlements at some time in the future, with transition arrangements for rapidly growing developing countries, was most likely to serve as the basis of global agreement.

Third, it would secure a level playing field for investment and trade in emissions-intensive goods, either through trade in emissions entitlements across countries, or through agreed rules on assistance to trade-exposed industries.

Fourth, it would embody a commitment by high-income countries to allocate a minimum sum, related to national income, to public financial support for research, development and commercialisation of new low-emissions technologies.

Fifth, it would contain commitments from developed countries to provide funding to support the climate change adaptation efforts of developing countries.

The Review (Chapters 8 to 10) suggested a possible basis for agreement in each of these areas.

The international community is closer to a basis for agreement in some of these areas than others.

On the first, there is widespread rhetorical support for strong mitigation, built around securing emissions concentrations of 450 ppm or below. It may not be difficult to secure an agreement on an ambitious mitigation objective. The problem is that without agreement on allocations across countries that add up to the desired total, it is only rhetorical.

The second area in which agreement is required is crucial. This is a technically complex matter with large ramifications for the distribution of income in and between countries. It will take time to build an understanding across all countries. It is unlikely that there will be effective agreement unless heads of major governments have put in place a process for sorting through the possibilities, with reporting times well in advance of the Copenhagen meeting. This process has not yet begun. The position of the United States is crucial, and the late start in that country and the pressures of recession may make it difficult for it to play a leading role in time for agreement in December 2009. Chinese participation is likely only after a strong lead from the United States, and other developing countries later still. Time is running out.

The third area in which agreement is required has only recently been given anything like the attention it requires. As soon as countries begin to take strong action to reduce emissions, the possibility of emissions prices in the home country exceeding that in some or other competitor becomes a source of agitation for assistance. Each country which has or is contemplating strong mitigation is developing its own approach to assistance. The inconsistencies in approach create infinite opportunities to argue for pressure for increased assistance in every country. The unedifying Australian discussion of allocation of free permits has its analogues everywhere. In the United States, it is manifested in discussion of measures to penalise imports of countries whose industries are not exposed to comparable emissions pricing.

This is a matter on which international rules are crucial. The absence of international rules would become an argument against strong mitigation in every country. It is likely to lead to systematic exclusion of many of the most emissions-intensive industries from constraints being applied elsewhere in the economy. It is likely to corrode the multilateral trading system.

The problem would not arise if there were comprehensive allocation of emissions rights and trade in those rights. This would establish comparable emissions pricing in all participating countries and a level playing field for competition in the emissions-intensive industries.

Pending agreement on allocating emissions rights as a basis for global trading of permits, the solution is a principled approach to assistance to trade-exposed, emissions-intensive industries in all countries. The principled approach is defined in Section 14.5 of the Review (see also Garnaut 2010). Each country would assess the effects on global prices of other countries not having comparable emissions pricing, and compensate domestic producers for divergence of 'shadow' from actual pricing.

The global solution prior to comprehensive emissions pricing and trade is to establish an international entity to assess the carbon price that would correspond to universal carbon pricing at various rates, and to allow support to enterprises in each country to the extent that there was divergence between the current international price, and the price that would rule if that price applied in all countries.

The fourth requirement in an effective global mitigation agreement is for a minimum commitment by high-income countries to public fiscal support for research, development and commercialisation of new, low-emissions technologies. This is one aspect of agreement that has become easier with recession and fiscal expansion across developed countries. The unilateral commitments on technology of the Obama Administration are broadly in line with, if still below, what is required from the United States within a global agreement. This is one dimension of mitigation policy that is both consistent with recovery policies, and widely perceived to be so. Now is a good time to lock in global agreement on one of the pillars for an ambitious global mitigation effort.

The requirement of adaptation assistance for developing countries, especially those with low incomes and most vulnerable to climate change, has a particular history that makes it a condition for the participation of some developing countries in a comprehensive global mitigation effort. Adaptation assistance would often be administered through a development assistance budget and agency. Adaptation assistance could be a means through which many countries met stated commitments to higher levels of development assistance.

Global recession and the Australian policy discussion

The whole reason for Australian action is to encourage the emergence of an effective global mitigation effort. The first test that Australian policy must pass is that it does that well.

The Garnaut Climate Change Review makes the case that it is in Australia's national interest to secure the strongest possible global mitigation agreement. This follows from Australia being the developed country likely to be most severely damaged by unmitigated or weakly mitigated climate change, and from analysis demonstrating that the benefits to Australia of strong global action, with Australia playing its full proportionate part, outweigh the costs. This case was accepted by the Government in the White Paper (Department of Climate Change 2008).

It is important that we indicate that Australia is prepared to play its full proportionate part in an ambitious global agreement on mitigation. The Government's White Paper in December 2008 proposes targets that indicate willingness to do our proportionate part in an international agreement of considerable but not of high ambition. Reduction of emissions to 15% below 2000 levels would correspond to our share in a global agreement to hold emissions at somewhere below 550 ppm but above 500 ppm.

It must be said that an effective agreement along these lines would be a considerable step forward.

The maximum reduction of 15% would not, however, allow us to play our proportionate part in an agreement to secure emissions concentrations at or below 450 ppm which, if it were feasible, would correspond more closely to the Australian national interest. The Government's change of position to keep alive the possibility of a 25% reduction on 2000 levels by 2050 is therefore greatly to be welcomed. Announcement of our willingness to participate in an ambitious agreement encourages the emergence of such an outcome. An ambitious outcome might just become possible, as the dynamics of United States relations with China unfold in the initial stages of the Obama Presidency.

An effective international agreement on the allocation of emissions entitlements with high levels of ambition may not be feasible by December 2009. World leaders may decide to lock down agreement on some issues in December, but that delay in agreement on entitlements allocations across countries is better than compromised ambition.

Now is a good time to lock in place international agreement on a low-emissions technology commitment. Australia is in the process of greatly increasing its support for research on carbon capture and storage from fossil fuel combustion. The expansion of this commitment to greatly increased funding for research, development and commercialisation of low-emissions technologies in which Australia has a national interest and comparative advantage has a logical place in recovery strategies. These areas would include biosequestration of several kinds, which is potentially transformative in both the cost and the potential extent of emissions reduction within Australia. Expansion of Australian effort now on the new technologies would allow Australia to play a part in movement towards the necessary global commitments.

Why go further at this difficult time, than to announce our willingness to play our full part in an ambitious global agreement, should one be reached, and to increase the commitment to the emergence of superior low-emissions technology? For the other necessary elements in an effective mitigation policy, and especially for the pricing of carbon, why not wait until the financial and economic storms have passed, and we know the outcome of the Copenhagen and subsequent meetings?

In particular, why do we need to put a price on carbon now, when we seem to be on track now to meet our 2012 Kyoto targets?

These are all good questions.

There are several good reasons for locking in place now the structure of an emissions trading scheme (ETS). A properly designed scheme is likely to deliver emissions reductions of specified extent with greater certainty and at lower cost than the alternatives. This is especially so when the opportunities for international trade in permits are considered. The ETS is a major institutional development. Time is required to iron out inevitable imperfections, before it is called upon to carry a heavy load of emissions reductions.

Good work is being done to put an ETS in place. There are advantages in making full use of that work. Making a start in 2011 makes it more likely that Australia will have an effective instrument for reducing emissions at relatively low cost when it is required, from 2013.

The Review recommended early introduction of the scheme, but allowing market participants to buy permits at a fixed price during the remainder of the Kyoto period, to the end of 2012. That would allow the regulatory agencies and market participants to become familiar with compliance and monitoring processes. Market participants would be assured of a low permit price for and beyond what we all hope will be the full duration of the crisis. Financial markets would not have to carry a heavy load until they had recovered from the stresses of 2008 and its aftermath.

The greatest difficulties of implementation of the ETS relate to 'compensation' for trade-exposed industries. The White Paper says that the principled approach to assistance put forward by the Review is correct in principle. It has been said, however, that the principled approach would be too difficult for the Australian authorities to administer. It is clear by now, if it was not clear before, that the *ad hoc* approach favoured by the White Paper is not plain sailing.

A low fixed price for permits to the end of 2012 would reduce the costs of distortions from *ad hoc* arrangements for compensating trade-exposed industries during the Kyoto period. The fixed price period would provide time for the introduction of a principled approach from 2013, whether on a national or an international basis. It would provide time for development of an approach within the WTO, if there were the will amongst major Governments to do so.

To delay introduction of the scheme altogether until it was needed to secure large reductions in emissions would carry high risks. The presence of the scheme would signal that, when the time arrived, the pricing of emissions through an ETS was to be the principal instrument of Australian mitigation policy.

Its absence would leave a vacuum. The high community interest in climate mitigation would make it certain that the vacuum would be filled. It is likely to be filled by manifold policy interventions, together with less potential for emissions reduction, and at much greater cost, than an ETS. This should be of concern not only for people who recognise that economic and environmental efficiency are complementary, but also for those who are interested solely in either economic efficiency or climate change mitigation.

The other type of reason for pressing ahead with introduction of an Australian ETS is that time is running out for a global agreement with prospects for holding risks of dangerous climate change to acceptable levels.

Pressing ahead with the ETS would provide a signal to the international community that Australia was following through on the commitments made in Bali in December 2008. This would be helpful to global movement towards agreement in the approach to Copenhagen.

To go beyond amendments to improve the proposed ETS, and to defeat it comprehensively, would have global implications. It would be noticed in the United States debate about an ETS. It would raise doubts about Australia's capacity to join a strong international mitigation effort. This would be especially important amongst developing countries who heard the developed countries including Australia make strong commitments to lead on mitigation, first at Rio de Janeiro in 1992, then at Kyoto in 1997, and most recently December 2008 in Bali. It would set back progress towards a global agreement.

A time to conserve what is good

I referred in the Climate Change Review to things falling apart when society receives a shock that is too large for human institutions to cope.

That is the ultimately conservative case both for strong climate change mitigation, and for strong action to restore high employment and rising incomes.

The financial and economic crisis and the now urgent challenge of climate change make this a time for careful analysis of policy choice on climate change and the economic crisis, and of their interaction with each other. It is a time for strong involvement in the policy process of a large centre of the polity, whose involvement is motivated by concern for the public interest.

Alas, it has recently been a time when the Australian discussion has been claimed dispro-portionately by the private interest, the ignorant, the myopic and the excessive.

> Turning and turning in the widening gyre
> The falcon cannot hear the falconer
> Things fall apart; the centre cannot hold.
>
> – WB Yeats 'The Second Coming'

This is an edited transcript of a speech given at the GREENHOUSE 2009 conference.

References

Department of Climate Change (2008) *Carbon Pollution Reduction Scheme: Australia's Low Pollution Future White Paper*. Commonwealth of Australia.

Garnaut R (2008) *The Garnaut Climate Change Review*. Cambridge University Press, Melbourne. <www.garnautreview.org.au>

Garnaut R (2010) Climate change and the Australian agricultural and resource industries. *The Australian Journal of Agricultural and Resource Economics* **54**(1), 9–25. doi:10.1111/j.1467-8489. 2009.00475.x.

Garnaut R, Howes S, Jotzo F and Sheehan P (2008) Emissions in the Platinum Age: the implications of rapid development for climate change mitigation. *Oxford Review of Economic Policy* **24**, 377–401.

Garnaut R, Howes S, Jotzo F and Sheehan P (2009) The implications of rapid development for emissions and climate-change mitigation. In: *The Economics and Policy of Climate Change*. (Eds D Helm and C Hepburn), Oxford University Press, Oxford.

Garnaut R with Llewellyn-Smith D (2009) *The Great Crash of 2008*. Melbourne University Publishing, Melbourne.

Stern N (2007) *The Economics of Climate Change: The Stern Review*. Cambridge University Press, Cambridge.

Paul Holper

Part 1

Climate change science

Abstract

The Commonwealth Government began dedicated funding for climate change research in 1989 in response to concerns about the likely impact of increases in atmospheric concentrations of greenhouse gases. Over the past 20 years, CSIRO and Bureau of Meteorology researchers have made internationally recognised advances in atmospheric and oceanic science, and in our understanding of biospheric changes and modifications in the carbon cycle. They have developed sophisticated climate models and regularly released projections on the way in which Australia's climate is likely to change in future decades.

Introduction

It was abundantly clear during the 1980s that atmospheric carbon dioxide concentrations were rising. Since the inception in 1976 of the Cape Grim Baseline Air Pollution Station in north-western Tasmania, the measurements of pristine air mirrored what US researchers had found atop Mauna Loa in Hawaii since the 1950s: each year the atmosphere contained more greenhouse gas. But what were the likely consequences of this change?

Here was a potential environmental challenge with major implications for the southern hemisphere. CSIRO, under the leadership of Dr Graeme Pearman, joined forces with the Commission for the Future to organise a multidisciplinary conference to draw attention to the matter. GREENHOUSE 87 spawned a seminal publication, *Greenhouse: Planning for Climate Change* (Pearman 1987), which challenged Australia's pre-eminent scientists to consider the science and consequent impacts of the 'enhanced greenhouse effect'. Two years later, the Commonwealth Government granted CSIRO and the Bureau of Meteorology funds to undertake research into the phenomenon.

In 2009, the Australian Climate Change Science Program (ACCSP) entered its 20th year of continuous Commonwealth Government support. Since its inception, the Program has been the catalyst for a national effort involving CSIRO, Bureau of Meteorology and leading universities to track, explain and project climate change.

ACCSP researchers have monitored the air, probed the oceans and explored the interaction between the landscape and climate. They have used instruments on satellites to measure vegetation cover over the continent, behaviour of dust and aerosols, and changes to sea-surface temperatures. Scientists have taken computer models of atmospheric behaviour and developed them into sophisticated tools that simulate global climate, from the depths of the oceans to the atmosphere; from the biosphere to the ice-covered poles.

The ACCSP has produced projections showing how Australia's climate is likely to change in the future four times, each set of projections more comprehensive and detailed than the last. In 2007, there were the first probabilistic projections, revealing the likelihood of particular increases in temperature or changes in rainfall.

Through its focus on the southern hemisphere, the ACCSP has contributed significantly to understanding of El Niño, the Antarctic Circumpolar Current, the Indian Ocean Dipole and many other large-scale phenomena that influence global climate.

The Department of Climate Change and its predecessor Commonwealth agencies have supported research and worked closely with the scientists to communicate findings to decision makers, industry and the community, providing the critical underpinnings for the development of mitigation and adaptation policy, business planning and community action.

The contribution of ACCSP researchers to the global understanding of the Earth's climate system and climate change has been recognised nationally and internationally, through hundreds of insightful papers in prestigious journals such as *Nature* and *Science*, through dozens of individual and team awards and medals, and through many ACCSP scientists contributing to the Intergovernmental Panel on Climate Change (IPCC). In 2007, the IPCC shared the Nobel Peace Prize.

Research highlights

In the beginning

In the 1980s, climate modelling was in its infancy. CSIRO's first Australian experiment to examine climate change was undertaken using a four-level model of the atmosphere joined to a primitive 'slab' model of the oceans. Researchers undertook an initial simulation using existing atmospheric concentrations of carbon dioxide, then repeated the simulation with doubled concentrations of the gas. The result: a global mean warming of 4°C. Meanwhile, the Bureau of Meteorology's nine-level model simulated a warming of 2°C.

CSIRO oceanographers developed a model of sea level rise due to thermal expansion of the oceans. Model simulated sea level was consistent with the 12 cm rise observed over the previous 80 years. The projection was for an increase of 20 cm over the subsequent 30 years.

Colleagues undertook a research voyage south of Australia to measure the physics, chemistry and biological productivity of the oceans. They found evidence that this region is an oceanic sink, or net absorber, of atmospheric carbon dioxide. Ice-core research revealed a doubling of atmospheric methane concentrations over the previous 150 years. Plant scientists studied wheat fields to determine the likely impacts of climate change and increasing carbon dioxide concentrations on water use efficiency.

Tracking gas

In 1989, carbon dioxide concentrations were approaching 350 ppm; today they are over 380 ppm.

The Cape Grim Baseline Air Pollution Station (a joint program of the Bureau of Meteorology and CSIRO) has tracked these changes over the years, also providing air samples that yield information about southern hemispheric sources and sinks of greenhouse gases through measurements of various isotopes.

In 1990, the influx of Commonwealth funding allowed CSIRO to open GASLAB (Global Atmospheric Sampling Laboratory) at its Aspendale site south of Melbourne. The laboratory's mass spectrometer and sensitive gas chromatographs were soon analysing air from Cape Grim as well as from many other sites worldwide. CSIRO had become a major importer of air!

Across the corridor was ICELAB (Ice Core Extraction Laboratory). Here, air from ice cores supplied by the Australian Antarctic Division revealed past changes to the composition of the atmosphere.

Modelling

Domestic climate modelling capacity is important as Australia is particularly vulnerable to the risks of climate change and year-to-year climate variability. We already live with climate extremes such as heatwaves, tropical cyclones and bushfires; climate change is likely to exacerbate these hazards. Researchers need to be able to simulate the way in which Australia and our region are likely to be affected, with a model specifically designed for our needs. National modelling also makes a powerful contribution to international assessments by the IPCC.

By 2005, it had become obvious that neither CSIRO nor the Bureau of Meteorology alone had the resources to continue to develop a world-class climate model. So the two agencies, along with universities, joined to develop the Australian Community Climate and Earth System Simulator (ACCESS).

ACCESS would have to provide the full range of weather and climate needs, from daily weather forecasts to climate change simulations for decades and centuries hence. In a departure from home-developed components, the Bureau and CSIRO elected to adopt a number of modules from agencies such as the United Kingdom Meteorological Office.

Today, Australian researchers are working on key aspects of identifying, understanding and modelling climate and weather drivers in the southern hemisphere, such as prediction of El Niño events, the Indian Ocean Dipole and simulation of the Southern Ocean, to feed into ACCESS. They are also building CABLE, a new land surface scheme to ensure that Australian vegetation is well represented, and a global ocean and sea ice model (AusCOM) to capture the major features of southern hemisphere ocean circulation.

Tracking the oceans

Researchers have made substantial progress in understanding the behaviour of the oceans surrounding the continent and their role in the climate system. These advances have come from extensive observations from ships, moorings, drifting robotic floats and satellites, and from our ability to simulate the interactions between the atmosphere, ocean, sea ice and biogeochemistry.

As in so many fields, Australian researchers have been major contributors to international science efforts. The Argo program, for example, is a major international collaboration to observe the world's oceans. The data collected also improves our ability to forecast climate and ocean conditions. Argo consists of a global array of robotic floats that drift at depths of between one and two kilometres. Every 10 days, each float slowly ascends to the surface, regularly measuring temperature and salinity. This data is transmitted to satellites along with the float's position.

Throughout the 1990s, Australia contributed to the first global survey as part of the World Ocean Circulation Experiment, producing the most definitive estimates of the oceans' role in transporting heat and freshwater through our climate system, and permitting the first global ocean inventories of nutrients and carbon.

Interactions between the atmosphere, the oceans and sea ice in the Southern Ocean result in formation of water masses that play an important role in the global overturning circulation, a process that draws surface waters down to the ocean's abyss near Antarctica and deep waters up to the surface. CSIRO scientists have helped quantify how the Southern Ocean 'ventilates', or adds oxygen to, large volumes of the global ocean, and in understanding the important role of the Southern Ocean in acting as a carbon sink.

Researchers have examined how processes such as the El Niño–Southern Oscillation and fluctuations in sea-surface temperature across the Indian Ocean affect Australian climate variability and change. Oceanographers have measured the Antarctic Circumpolar Current and the Indonesian Throughflow, a movement of warm, low salinity waters from the tropical Western Pacific through the Indonesian seas into the Indian Ocean.

CSIRO researchers have also improved estimates of changes to ocean heat content, and improved the representation of ocean eddies in climate models.

Sea level

The National Tidal Centre within the Bureau of Meteorology undertakes monitoring and analysis for establishing trends in sea level.

Modelling and measurements have identified an acceleration of global sea level rise and permitted a more accurate view of what has driven sea level rise over the past 50 years. A team of Australian and US climate researchers found that the world's oceans warmed and rose at a rate 50% greater in the last four decades of the 20th century than documented by the 2007 IPCC Report.

New models have improved projections of sea level rise and of the frequency and magnitude of high sea level events, known as storm surges. The projections indicate that a sea level rise of 0.5 metres could result in current one-in-100 year events occurring more than once a year.

Climate projections

By building on results from climate models as well as historical and palaeoclimate records, researchers in the late 1980s pioneered development of scenarios of the likely impact of climate change. This provided the basis for many studies undertaken by groups throughout Australia. To generate regional information from the coarser scale provided by global models, researchers developed downscaling techniques, such as embedding a localised model for the area of interest. Another research team created an innovative 'stretched grid' global model, in which grid spacing is small in regions of interest and large elsewhere.

The CSIRO climate impact group grew rapidly to fulfil the rising demand for information, releasing climate 'scenarios' for Australia in 1992, 1996 and 2001.

The latest set of projections, released in 2007, emerged from collaboration between CSIRO and the Bureau of Meteorology and represent the most comprehensive assessment to-date of Australia's future climate. *Climate Change in Australia* (CSIRO and Bureau of Meteorology 2007) provides details on observed climate change and its likely causes, and projections of changes in temperature, rainfall and other aspects of climate for the coming decades.

By 2030, temperatures over Australia will rise by about 1°C compared with recent decades. The amount of warming later this century will depend on the rate of greenhouse gas emissions. If emissions are low, warming by around the year 2070 will be between 1°C and 2.5°C, with a best estimate of 1.8°C. However, with high emissions, the warming range will be 2.2°C to 5°C, with a best estimate of 3.4°C.

For the first time, the projections presented probabilities of likely changes. For example, with high greenhouse emissions, the chance of warming greater than 4°C by 2070 is around 10% in most coastal areas and 20–50% inland.

Future decreases in rainfall are likely in southern areas during winter, in southern and eastern areas during spring, and in south-west Western Australia during autumn, compared with conditions over the past century. Droughts are likely to become more frequent, particularly in the south-west of Australia, high fire danger weather is likely to increase in the south-east, and sea levels will continue to rise.

Details are available at www.climatechangeinaustralia.gov.au.

Terrestrial changes

We live in a unique land. Our terrestrial ecosystems are markedly different from those in North America and Europe. In 2000, CSIRO built a 70 m high tower at Tumbarumba in the Bago State Forest in south-eastern New South Wales. High above the tree tops, instruments measure the movement of water vapour, carbon dioxide and heat in and out of the forest.

Researchers discovered that the timing and amount of rainfall along with the prevailing humidity dictate how rapidly the eucalypts grow. In extreme conditions, such as the drought years of 2002–2004, the trees and plants under insect attack reverted from their usual role of absorbing carbon dioxide from the air to become a source of the greenhouse gas.

Measurements from the Tumbarumba tower have enabled a complete carbon budget assessment for a productive forest, as well as demonstrating the large inter-annual variability in uptake of carbon. With results from another tower at Virginia Park in the northern Queensland savanna, researchers have learnt much about interactions between terrestrial carbon and water cycles and the climate, and improved the modelling of these processes.

Reviews

Roy Green and Don MacRae (2003) were complimentary when they reviewed the ACCSP: 'Quite profound progress has been made since 1989, both in Australia and internationally, in our observation and understanding of the climate system, leading to improved climate projections through the development of more sophisticated and representative climate models.' The review noted the role of the research in helping 'progress in mitigation technologies, in assessment and monitoring, and in understanding the likely effects of climate change on our environment, our economy and our society'.

Four years later (2007), eminent US researcher Susan Solomon and Will Steffen from ANU found that the program 'plays a more significant role in the overall national climate change research effort than the size of its budget might indicate. The program has acted as catalyst for a much larger effort within the participating research organisations, and also as an institutional integrator that facilitates collaboration among scientists from a range of institutional bases'.

The Solomon and Steffen review continued:

> The (program) has produced an impressive body of work that has clearly been state-of-the-art and world class: i.e., of the highest impact on climate change science, on a par with the best achievements obtained by the top research groups internationally. The success owes much to the involvement of a cadre of outstanding researchers who are well-recognised for excellence. Further, Australian climate scientists have managed to skilfully address a suite of uniquely Australian climate change science topics, while at the same time exploiting and indeed pioneering the ways in which many of these lead to advances in the broader understanding of global climate change.

Communication

Research without communication is pointless. A major communication event, the national GREENHOUSE 87 conference, helped galvanise initial support for climate change research in Australia. During its 20 years, the research program has actively encouraged its scientists to explain their findings and the significance of them to the public, to government and industry.

The program has issued more than 100 media releases, published dozens of flyers and brochures, and arranged many media conferences, workshops, symposia and conferences. Scientists

4. TROPICAL AUSTRALIA AND THE AUSTRALIAN MONSOON: GENERAL ASSESSMENT AND PROJECTED CHANGES

Aurel Moise and Robert Colman

Abstract

The Australian monsoon is a fundamental component of southern hemisphere summer circulation, and dominates rainfall distributions over northern Australia and adjacent regions. Changes to the Australian monsoon over the coming century could have profound consequences for northern Australia and adjacent regions, affecting rainfall totals, distribution or intensity. This chapter examines the ability of the current generation of models to simulate the Australian monsoon, including basic temperature, pressure, wind and precipitation patterns and variation. The variability of models on a range of time scales is also assessed, including interannual variability. We find that while there are some deficiencies in simulating 20th century monthly climate means (of rainfall in particular), some of the large-scale features such as the zonal winds are quite reasonably simulated. Assessing projected changes in the Australian monsoon under enhanced greenhouse conditions is problematic, because the signal (if any) seems to be very weak. The ensemble mean model results indicate a possible later retreat of the monsoon, in particular over the north-western part of Australia.

Introduction

Large uncertainties remain in possible climate changes caused by increasing greenhouse gases, particularly at regional levels (IPCC 2007a; CSIRO and Bureau of Meteorology 2007). The Asian-Australian monsoon comprises an important large-scale component of the climate system, as well as being the major source of regional seasonal rainfall variation, and changes in the monsoon may have major impacts. The Australian component of monsoon has been little studied, and although populations are much smaller, regional impacts may also be large particularly on vulnerable indigenous populations and on ecosystems (IPCC 2007b). This is especially the case since the Australian monsoon is the dominating factor in tropical Australian rainfall (see, for example, McBride 1998), and uncertainties in possible changes to the mean monsoon or to monsoon variability remain key uncertainties in Australian regional climate change projection (IPCC 2007a, 2007b). This study will examine how well large-scale monsoon features for the current climate are simulated by state-of-the-art Coupled Global Climate Models (CGCMs), as represented by the models that took part in the World Climate Research Program's (WCRP's) Coupled Model Intercomparison Project phase 3 (CMIP3) (Meehl *et al.* 2007). We will also consider projected monsoon changes for the following century. Although very few studies have focused specifically on evaluation of climate model representation of Australian monsoon circulations, a significant number have considered aspects of the broader Australian climate, particularly relating to temperatures and precipitation (CSIRO and Bureau of Meteorology 2007). As part of the Atmospheric Model Intercomparison Project (AMIP2), Zhang *et al.* (2002) analysed 16 atmospheric GCMs, finding significant model deficiencies in rainfall and surface temperature seasonality. Moise *et al.* (2005), examining 18 CGCMs taking part in the Coupled Model Intercomparison Project phase 2 (CMIP2), found agreement with the gross spatial patterns of austral summer rainfall (DJF), but with significant model errors in simulating the intensity and location of heavy monsoonal rainfall in the north and eastern parts of the continent. Seasonal surface temperature variations were reasonably reproduced, but with biases of around 2 to 4°C. Wang *et al.* (2004) found that atmospheric GCMs forced by sea surface temperature (SST) anomalies skilfully simulated northern Australian monsoon related zonal wind variability for the period 1996–1998. A much earlier study by Suppiah (1995) showed that one GCM was able to simulate many of the observed large-scale aspects of Australian monsoon circulation, such as the seasonality of rainfall and temperature, and the location of the low level 'monsoon trough', although it was much weaker than observations in its representation of the magnitudes of winds and precipitation.

Model and validation datasets

Models considered were those submitted to the WCRP's CMIP3 (Meehl *et al.* 2007). Monthly results were extracted from the 20th century coupled runs (20C3M) for the period 1980–1999 for analysis of mean fields, and the longer period, 1950–1999, for interannual variability (see AchutaRao *et al.* 2007 for a summary of model forcing). A single realisation was selected for each model where multiple realisations were available. Variability of climate between different ensemble members for individual models, however, was investigated and is addressed below. Australian daily mean surface temperatures were calculated by averaging maximum and minimum temperatures from the high-quality observational data sets compiled by the National Climate Centre of the Australian Bureau of Meteorology (Jones and Trewin 2000). Precipitation observations over Australia were taken from the 0.25° monthly high quality gridded dataset (based on gauge data) developed by the Australian Bureau of Meteorology (Jones and Weymouth 1997). For inclusion of surrounding oceans (as in the bias plots below) the Climate Prediction Center Merged Analysis of Precipitation (CMAP) blended gauge/satellite data set was used (Xie and Arkin 1997). Winds and surface pressure were taken from monthly means from the European Centre for Medium Range Weather Forecasting (ECMWF) 40 year reanalysis (ERA40) (Uppala *et al.* 2005).

The model-simulated monsoon climate

Seasonal cycles over tropical Australia

This section will focus on the seasonal variation of important climate features associated with the monsoon, and their representation by models. Firstly, large-scale averages over northern Australia (120–150°E, 10–20°S, see Figure 4.1) of surface air temperature (TAS), mean sea level pressure (MSLP) and precipitation as well as 850 hectopascal (hPa) and 200 hPa zonal winds are shown in Figure 4.2. Characteristic features of the observed monsoon are a strong seasonal cycle of precipitation, with virtually none falling during winter months, increasing rapidly in October and November, to a peak of around 7 mm per

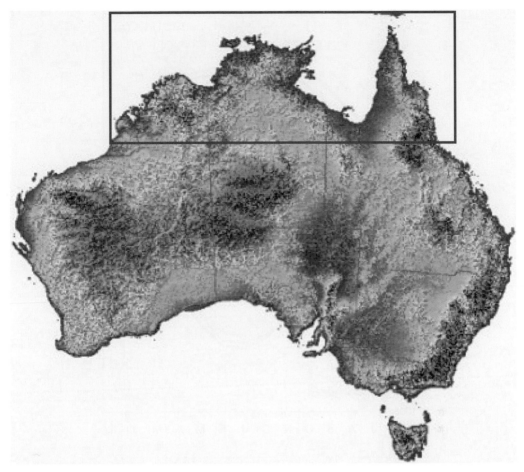

Figure 4.1: Map showing the tropical area of Australia (10–20°S, 110–150°E) considered in this chapter.

day on average during February. Rapid falloff occurs in March/April. The mean cycle, of course, disguises a large amount of interannual and intraseasonal variability (Hendon and Liebmann 1990 a, b; Holland 1986; Drosdowsky 1996), and some of these features are discussed below. The MSLP change from summer to winter is around 7 hPa, with minimum climatological values occurring in January/February, associated with maximum southward extent of the monsoon trough (see, for example, McBride 1998). Note that averaging of MSLP here includes ocean points, so as to reduce the impact on the mean of the (shallow) surface 'heat low', typically located in the north-east of Western Australia, which occurs commonly over summer months (Gunn *et al.* 1989). Associated with the motion of the monsoon trough, there is seasonal reversal of both lower and upper winds: low level winter trade wind easterlies are replaced by westerlies peaking at around 2 metres per second (m/s) in January and February, and extending up to 500 hPa. Upper level easterlies also occur for the peak monsoon period. Temperature displays a seasonal variation of around 10°C, with the peak notably occurring early in the wet season in November, followed by a slow decline until April, with more rapid decline thereafter. This is consistent with maximum warming occurring under relatively clear skies prior to the main onset of the monsoon (with subsequent increases in cloudiness and precipitation).

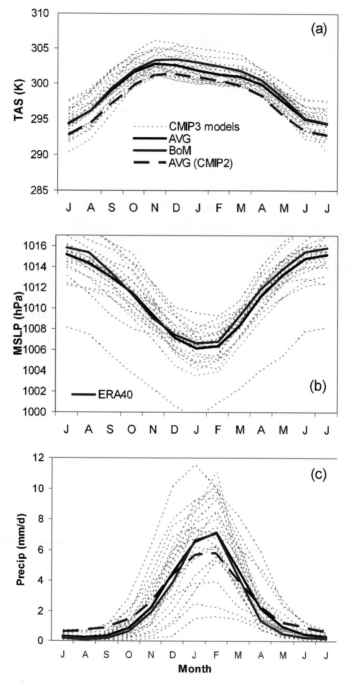

Figure 4.2: Seasonal variation, averaged over 120–150°E, 10–20°S for (a) daily mean surface air temperature (K), (b) mean sea level pressure (hPa) and (c) precipitation (mm/day), averaged over the period 1980–1999. Individual CMIP3 models are shown as dashed grey lines. Also shown are the multi-model means (black lines) and observational/reanalysis estimates (red). (a) and (c) show averages over land points only. Also shown are the averages calculated over 18 CMIP2 models (dashed black). Parts (d) and (e) show the 850 hPa and 200 hPa zonal winds averaged over the same region.

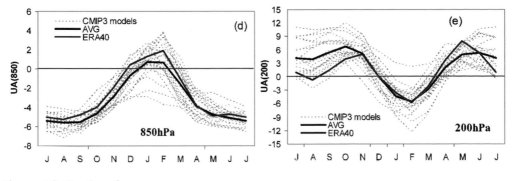

Figure 4.2: Continued

Mean climate and circulation

Firstly, large-scale patterns of surface temperature, MSLP and precipitation for austral summer (DJF) are compared for the average of the 24 CMIP3 models and the observational/reanalysis datasets (MSLP and temperature not shown here). Particularly for rainfall, substantial biases are found across the ensemble of models. Figure 4.3 shows that the ensemble mean model has a dry bias in tropical Australia and a wet bias further south, i.e. the north–south gradient in summer rainfall is not steep enough.

This is apparent also in the RMS error and spatial correlation statistics shown in Figure 4.4. However, a significant improvement over CMIP2 model performance is found for both statistics in the ensemble mean model during the Australian summer season. Because of the steep north–south gradient in tropical Australian rainfall, relatively small geographical mismatches will result in large underperformances in spatial correlation to the observations, as can be seen in Figure 4.4.

One of the distinct features of the Australian monsoon season is the reversal in the zonal winds at both low and high altitudes occurring during the onset of the monsoon. Figure 4.5

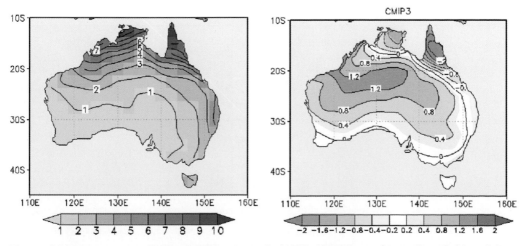

Figure 4.3: Mean summer (DJF) rainfall for Australia (1980–1999) from observations (left) and the CMIP3 ensemble mean model bias (right). Units are millimetres per day (mm/day).

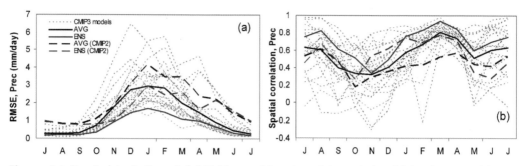

Figure 4.4: Tropical Australian rainfall RMS error (a) and spatial correlation (b) for the ensemble mean (red) and average (black) model from CMIP2 (long dashed) and CMIP3 (solid) inter-comparisons. The individual CMIP3 models are depicted in grey.

shows this reversal for the ERA40 reanalysis and CMIP3 ensemble mean model for the period 1980–1999. From the reanalysis, the mean monsoon onset date for the entire tropical region considered is around late November and the mean retreat date is in late February. The low-level westerlies during the monsoon season extend upward to around 500 hPa with easterlies at higher altitudes. While the overall height structure of the zonal wind is captured by the ensemble mean model, there are some deficiencies: the monsoon onset is too late (mid-December) and the retreat somewhat too early (mid-February); the low level westerlies are generally too weak; and they do not extend high enough (only to 700 hPa).

Interannual variability

The year-to-year variability in rainfall during the wet season is an important feature for the Australian monsoon. Figure 4.6 shows the October to April (ONDJFMA) rainfall variability across tropical Australia for the 20 year period 1980–1999 in the observed record and all CMIP3 models. Most models have a too-low interannual variability; however, in three models it is comparable to observations and too large in three others.

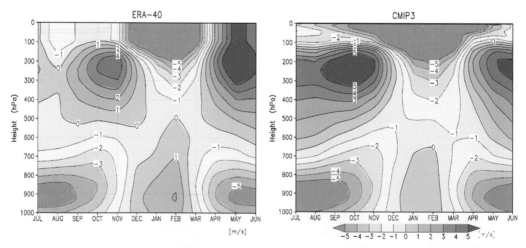

Figure 4.5: Seasonal cycle of mean zonal winds (m/s) vs. pressure level (hPa) for Australia (1980–1999) from ERA40 (left) and the CMIP3 ensemble mean model (right).

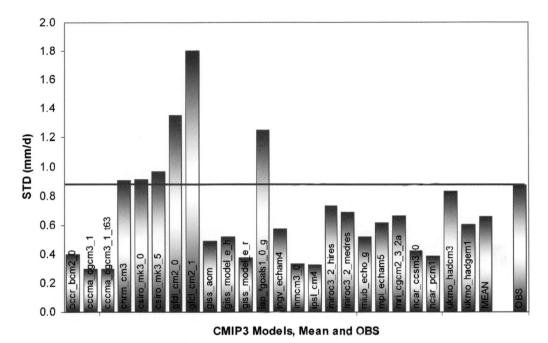

Figure 4.6: CMIP3 inter-annual variability of wet season (ONDJFMA) rainfall over tropical Australia for the models (shown with acronyms), the average of the model values (MEAN) and observations (OBS and red line). Units are mm/day.

Projected changes in tropical Australia

Change in mean climate

Recent results from the Global Carbon Project (2008) indicate that the global CO_2 emission pathway is currently tracking along the A2 scenario, the highest CO_2 emission SRES scenario considered in the IPCC Fourth Assessment Report (IPCC 2007a). This was highlighted at the International Scientific Congress on Climate Change (Copenhagen, 10–12 March 2009), where this point was mentioned in the first of the six key messages from the Congress (Richardson *et al.* 2009). All results in this section – except where noted – will refer to projected changes for the period 2080–2099 as simulated under the A2 scenario compared to the 20 year period 1980–1999.

Figure 4.7 shows the seasonal average changes in precipitation (in %) for the ensemble mean model. During the summer season (a), no significant change across tropical Australia can be seen. This is also the case for the pre-monsoon season from September to November (d), while decreases in rainfall are simulated across southern and western parts of the continent. Some increase in rainfall is seen in autumn from March to May (b) and winter from June to August (c) for tropical Australia.

The time series of rainfall shown in Figure 4.8 shows the large spread of model simulations for tropical Australia (land only points). For the entire 21st century, there is no trend for monthly mean rainfall changes in the ensemble mean model as well as the model spread (indicated by the width of the +/- one standard deviation (STD)).

Change in seasonal cycle of tropical rainfall and interannual variability

The lack of a signal in rainfall changes across tropical Australia does not extend into the deep tropics. Figure 4.9 shows the seasonal migration of monthly mean rain across the equator for

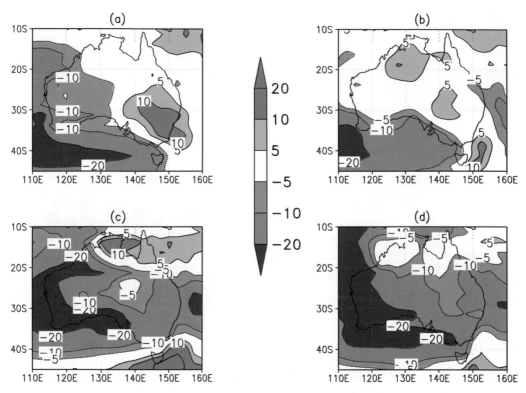

Figure 4.7: CMIP3 ensemble mean model changes in rainfall (units: %) for DJF (a), MAM (b), JJA (c) and SON (d). White areas indicate a change of less than +/- 5%.

the end of the 20th century (a) and 21st century (b). While these results are averaged over (100°–150°E), we have found very similar results for the global tropical belt.

The change in seasonal migration is shown in the two bottom plots (c) change in absolute units, and (d) change in %. Clearly visible is the main signal of increased rainfall in the deeper tropics (10% to 15% increase) with only little change south of 10° South during the Australian wet season (indicated by red box).

The model simulated interannual variability in rainfall over tropical Australia during the late 20th century was found to be too small for most models (Figure 4.6), and there is no coherent signal in how this variability would change under enhanced greenhouse conditions. Figure 4.10 shows that there are more models pointing toward reduced variability than not, but the overall average change is only in the order of −0.15 mm/d. Future work will assess the significance of this change.

Change in monsoon onset and duration

With results from the entire ensemble mean model of CMIP3 simulations pointing towards no changes in Australian tropical rainfall during the summer and only slightly reduced interannual variability, we investigated changes in the reversal of the zonal winds during the onset and demise of the monsoon over Australia.

There are large differences between models in their depiction of both onset date and monsoon duration. Here we present the ensemble mean model results (Figure 4.11) for the end of the 20th century (a), the 21st century (b), and their difference (c). The overall changes with regard to the monsoon season over Australia are fairly small: a slight intensification of the

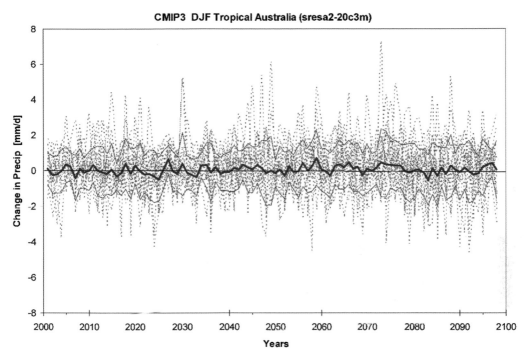

Figure 4.8: Time series of rainfall changes across tropical Australia (land points only) throughout the 21st century. Shown is the ensemble mean (thick blue line), +/- one STD (thin blue line), and all individual CMIP3 models (grey dashed lines).

westerlies at low levels combined with a slight weakening of the upper level easterlies. Furthermore, while the onset date (850 hPa wind reversal) seems to stay the same (mid-December), the retreat date is somewhat extended towards the end of March by the end of the 21st century.

This is also evident in the analysis of the monsoon shear line. It is known that a large fraction of tropical cyclones in the Australian region will develop along the shear line within several hundred kilometers of land (see, for example, McBride and Keenan 1982). Figure 4.12 shows the position of the monsoon shear line (defined via the 925 hPa zonal winds) for ERA40, simulated 20th century and 21st century for the months January (close to onset) and March (close to retreat). There is a large spread between the different model simulations (grey lines for the 20th century in Figure 4.12); however, the position of the ensemble mean model's shear line (light blue) is quite close to the ERA40 reanalysis data for the same time period. Under enhanced greenhouse gas conditions, the ensemble mean model shear line seem to only move slightly south over north-west Western Australia during the onset period, but more significantly so during the retreat period. This could indicate the later retreat of the Australian monsoon over the tropical north-west region of Australia. Further work is underway to investigate this issue in more detail.

Discussion and conclusions

This chapter describes some of the basic features of the Australian monsoon as simulated by the recent generation of coupled atmosphere ocean global circulation models. We find that while there are some deficiencies in simulating 20th century monthly climate means (of

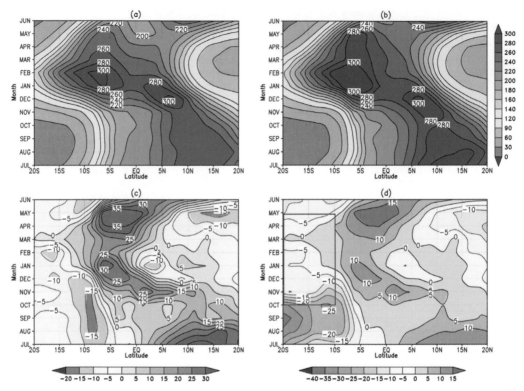

Figure 4.9: CMIP3 ensemble mean model seasonal migration of rainfall during 1980–1999 (a) and 2080–2099 (b) and the changes: (c) in units of (mm) and (d) in units (%). The red box indicates the area marked in Figure 4.1 during the wet season.

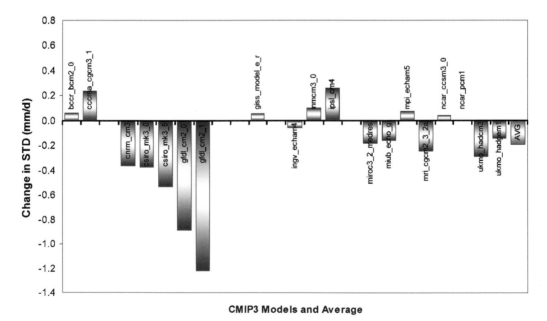

Figure 4.10: Change in tropical Australia rainfall interannual variability (in mm/day) during the wet season (ONDJFMA) of CMIP3 models.

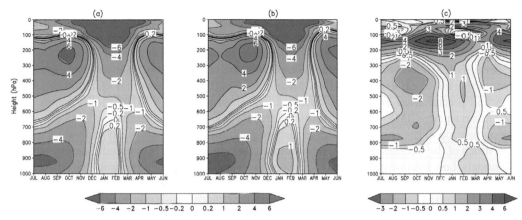

Figure 4.11: CMIP3 mean zonal winds versus height seasonality for tropical Australia over period 1980–1999 (a), 2080–2099 (b), and the difference (c). Units are (m/s).

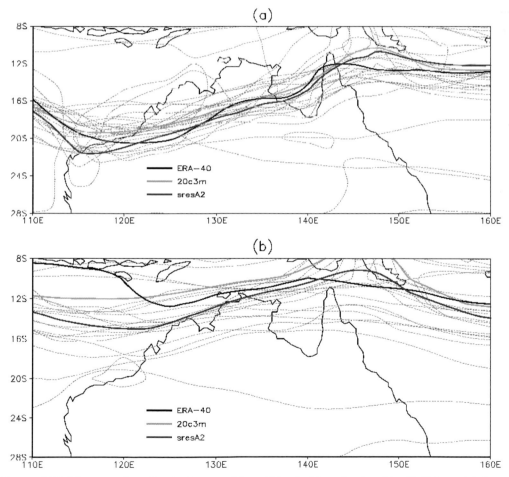

Figure 4.12: Monsoon shear line for January (a) and March (b) as simulated by CMIP3 models for the period 1980–1999 (light blue), 2080–2099 (red) and ERA40 for 1980–1999 (black). Individual CMIP3 models are shown in grey.

5. RECENT AND PROJECTED RAINFALL TRENDS IN SOUTH-WEST AUSTRALIA AND THE ASSOCIATED SHIFTS IN WEATHER SYSTEMS

Pandora Hope and Catherine Ganter

Abstract

Annual average rainfall declined in south-west Western Australia in the late 1960s and has not recovered since. A further shift to lower rainfall was identified at approximately the year 2000. The early decline was associated with a decrease in the frequency of occurrence of winter deep low-pressure systems crossing the region. The recent decline was not associated with a continuing decline in the number of deep low-pressure systems, but had a strong contribution from an increase in the number of weather systems with a high over the region. The recent rainfall decline had greater spatial extent than the decline in the late 1960s, in alignment with the decrease in rainfall being due to an increase in the number of high-pressure systems over the region, as highs generally influence a wider area of the south-west compared to low-pressure systems. These changes have occurred against the background of increasing mean sea level pressure (MSLP). Under the influence of enhanced levels of atmospheric greenhouse gases, climate models consistently show a signal of increased pressure and decreased rainfall by the end of the century. The magnitude of the mean changes in both rainfall and MSLP across the models for an intermediate future scenario is of the order of the changes seen since 2000 (2000 to 2007).

Introduction

Strong trends have been observed in annual rainfall across parts of Australia in the last 50 years (see, for example, www.bom.gov.au/cgi-bin/climate/change/trendmaps.cgi). One region with particularly strong trends is the south-west of Western Australia, where the well-documented rainfall decline in the late 1960s (Figure 5.1) caused major reductions in inflows to Perth's water storages (see, for example, www.watercorporation.com.au/D/dams_stream-flow.cfm). This decline inspired the initiation of the climate research program, the Indian Ocean Climate Initiative (IOCI). Further details on the IOCI program can be found at www.ioci.org.au.

The late 1960s decline in south-west Australian rainfall occurred primarily in May, June and July and is a very robust result, with a range of statistical tests finding a breakpoint at this time (IOCI 1999; Li *et al.* 2005; Ryan and Hope 2005, 2006; Hope *et al.* 2009). The decline was confined to the far south-west corner (particularly south-west of a line from 30°S, 115°E to 35°S, 120°E, known colloquially as the IOCI triangle, see Figure 5.1), (IOCI 2002; Hope and

Foster 2005). The decline was associated with an alteration of the large-scale atmosphere, which included a weakening of the subtropical jet in May and July and a reduction in the potential for storm development (Frederiksen and Frederiksen 2007; Frederiksen *et al.* 2009; Frederiksen *et al.* 2010). These large-scale changes are consistent with climate model projections under enhanced levels of atmospheric greenhouse gases (Cai *et al.* 2003; Fyfe 2003; Yin 2005). There was an associated reduction in the number of winter deep low-pressure systems crossing the region (Hope *et al.* 2006). Of less importance was an increase in the number of high-pressure systems over the region.

This chapter describes the ongoing work within the IOCI assessing the rainfall trends and variability since the late 1990s across the wider south-west. It then goes on to assess whether there have been further shifts in the frequency of storms or prevalence of high-pressure systems over the region associated with these rainfall trends. Observed changes are then compared with those projected for the end of the century.

Data

The monthly rainfall dataset used was the Australian Bureau of Meteorology's National Climate Centre (NCC) gridded monthly rainfall analyses from 1890 to 2008. The data is on a 0.25° by 0.25° latitude–longitude grid, based on all available station data that do not display gross errors available at any one time (Jones and Weymouth 1997). Temporal variations in the network can therefore have an impact on the trends as stations come in and out of the network. Monthly time series of area-averaged rainfall were created over the land points in the IOCI triangle – the area to the south-west of a line from the west coast at 30°S to 35°S, 120°E.

Monthly rainfall data from a number of high-quality stations across the south-west were also examined in this chapter. Stations were primarily selected from the high-quality monthly rainfall dataset developed by Lavery *et al.* (1997), with further testing for continuity of record in recent years. Where large spatial gaps existed between stations, stations with shorter record length than were considered by Lavery *et al.* (1997) were tested for quality, and those of high enough continuity of record and quality were also included. Graphs of the decadal variability of rainfall at these high-quality stations across the south-west are available at www.ioci.org.au.

A global MSLP dataset (HadSLP2r) from 1890 to 2008 on a 5° by 5° grid (Allan and Ansell 2006) was used to describe the MSLP trends for the south-west Australia region. The MSLP was averaged for a box encompassing 27.5° to 37.5°S and 112.5° to 122.5°E (see Figure 5.1). This dataset was preferred over the reanalyses as it is has a very long time series. Timbal and Hope (2008) compare HadSLP2r with reanalyses over this region for the period 1958–2002, and it has lower interannual variability, but was deemed satisfactory for the purposes required here.

Results

Observed recent rainfall changes

The rainfall decline in the late 1960s (Figure 5.1) has persisted to the present, and there is now evidence of a further rainfall decline, with a statistical break in the May to July rainfall totals for the south-west IOCI corner across 1999–2000 (Hope *et al.* 2009). A signature of the year-to-year variability of the rainfall since the late 1960s was a dearth of high rainfall years (Ryan and Hope 2005), and that signature has persisted into the most recent decline (Figure 5.2). The

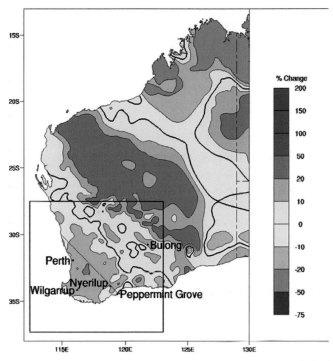

Figure 5.1: The percentage change in average 1969–1999 May to July gridded rainfall from the 1910–1968 mean. Also shown is the IOCI triangle that delimits the region considered for averaging south-west rainfall, the location of the high-quality rainfall stations referred to in the chapter and the square over which HadSLP2r MSLP data were averaged.

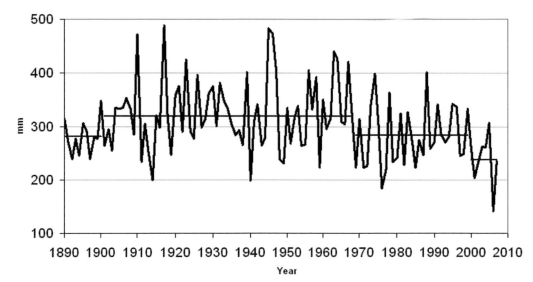

Figure 5.2: Time series of the monthly May to July total gridded rainfall data averaged over the IOCI triangle. Breakpoints in this series were found at around 1900 (up), 1968/1969 (down) and 1999/2000 (down) (Hope *et al.* 2009). The average totals for each of these epochs are marked on the graph and shown in Table 5.1.

Table 5.1   Average May to July total gridded rainfall for the IOCI triangle for relevant epochs

Epoch	Average May–July rainfall (mm)
1890–1900	281
1901–1968	319
1969–1999	284
2000–2007	238

average May to July rainfall for 2000–2007 is the lowest for each epoch delineated by the break points (Table 5.1), and also the lowest value for any eight year period within this record.

The decline in rainfall since 2000 is evident across a wider region than the previous decline, as can be seen from the percentage changes (average 2000 to 2008 rainfall minus the 1910–1968 average divided by the 1910–1968 mean, as a percentage (Figure 5.3), compared to the 1969–1999 percentage change shown in Figure 5.1). From a statistical perspective, more extreme values may be evident with shorter averaging periods such as the eight years for 2000–2007 compared to the 31 years for 1969–1999. These widespread reductions in rainfall, however, have real meaning for local water managers and communities. The season of the decline remains predominantly in May, June and July (Figure 5.4). The period since 2000 has totals across the IOCI triangle in May and June that are both lower than the rainfall in the 1890–1899 decade from which an upward step was found in the breakpoint analysis of Hope *et al.* (2009).

The response by month and decade at a number of high-quality stations across the wider south-west region shows a similar story to what was shown from the average over the far south-west. This is particularly true at stations within the IOCI triangle (e.g. Wilgarrup, Figure 5.1).

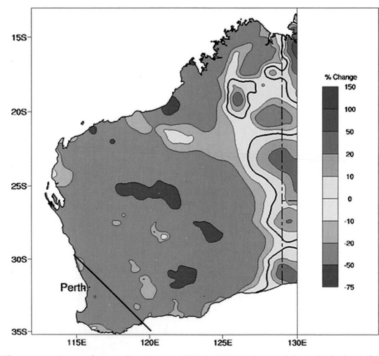

Figure 5.3: The percentage change in average 2000–2008 May to July gridded rainfall from the 1910–1968 mean.

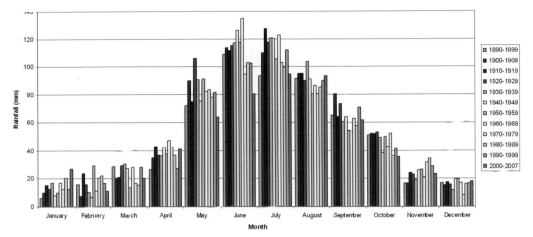

Figure 5.4: Decadal averages of rainfall in the IOCI triangle for each month.

Stations that showed minimal change or increases across the 1968/1969 break, for example along the south coast (Peppermint Grove, Figure 5.5) and further inland (Table 5.2), are now showing consistent declines, particularly in May and July. Table 5.2 shows the percentage changes across each of the recent breakpoints at a number of representative stations. For complete results from all stations deemed to be of high enough quality, that also provide an even spatial coverage across the wider south-west, refer to the IOCI website: www.ioci.org.au.

Shifts in the synoptic systems associated with the rainfall change

The synoptic patterns that describe the continuum of weather influencing south-west WA were identified with a self-organising map based upon 0:00 and 12:00 hour Coordinated Universal Time (UTC) mean sea level pressure each day of June and July from National Centers for Environmental Prediction (NCEP) and the National Center for Atmospheric Research (NCAR) reanalyses by Hope *et al.* (2006). It was found that the reduction in rainfall before and after 1975 was strongly associated with a reduction in the number of deep low-pressure systems that

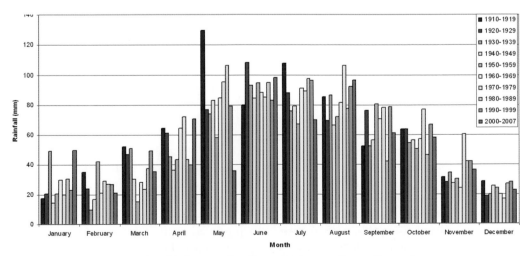

Figure 5.5: As Figure 5.4, but for Peppermint Grove station, on the south coast.

Table 5.2   The percentage change in May to July rainfall for each of the epochs since recent step-drops at a number of representative stations, compared to the 1910–1968 average

	1969–1999	2000–2007
Far south-west: Wilgarrup (Manjimup)	–25%	–44%
Wheat belt: Nyerilup (Mt Madden, Doongin Peak)	–14%	–35%
South coast: Peppermint Grove	+7%	–15%
Inland: Bulong (near Kalgoorlie)	+22%	–42%

bring extensive rainfall across the south-west. The time series of each synoptic type was extended through to 2008, and is shown in Figure 5.6. The types associated with extensive wet conditions across the south-west are shown along the bottom of the figure, while the driest types are along the top. Very wet years such as 1964 show a strong representation from a synoptic type associated with extensive wet conditions across the south-west (e.g. A5). Synoptic type A5, and the other types along the bottom of Figure 5.6 (e.g. B5, C5, D5) all display a deep trough to the south-west of Australia, with low pressures extending over the continent.

The percentage change over the period since 2000 (average of 2000–2008 minus the average of 1958–1999 divided by the average of 1958–1999) in the number of synoptic types associated with wet conditions (D4, A5, B5, C5, D5) was minimal (a reduction of 0.8%); however, the increase in synoptic types associated with a high over the region and extensive dry conditions (A1, B1, C1, B2, C2) was significant at 32%. This indicates that the recent changes have more to do with the persistence of high-pressure systems over the region, and less to do with a further decrease in the number of deep low-pressure systems. High-pressure systems influence a wider region than low-pressure systems and thus the recent rainfall declines might have been expected to have a greater spatial extent than the decline in the late 1960s, given the differing shifts in synoptic systems between the two periods. As noted above, the recent rainfall declines have indeed had a wider spatial extent than those in the late 1960s.

The mean sea level pressure over the region has increased through time. The May to July 1890–1968 average was 1017.8 hPa, while the average from 1969–1999 was 1018.3 hPa, an increase of 0.5 hPa. That increase has persisted through the rainfall decline since 2000. 2006 was a record-breaking year, with very low rainfall totals and mean sea level pressure values well above normal across southern Australia. Figure 5.7 shows the time series of May–July MSLP over the south-west, clearly showing that recent years were generally above average.

The designation of each day's MSLP pattern to a particular synoptic type in the self-organising map from Hope *et al.* (2006) is dependent on the magnitude of the MSLP as well as the pattern. Thus a background of increasing MSLP might be expected to produce the shifts to fewer low-pressure systems and more high-pressure systems described above. To test whether the shift in the frequency of particular synoptic types was due to the background pressure increases or an actual shift in the circulation, another method of defining high-pressure systems that depends on the local gradients rather than the absolute magnitude (Murray and Simmonds 1991) was investigated. The trends in the density per unit area of high-pressure systems calculated using this method have been produced at the Bureau of Meteorology's National Climate Centre (displayed at: www.bom.gov.au/silo/products/cli_chg/). In all seasons except summer, there is a clear upward trend in the number of systems (e.g. Figure 5.8

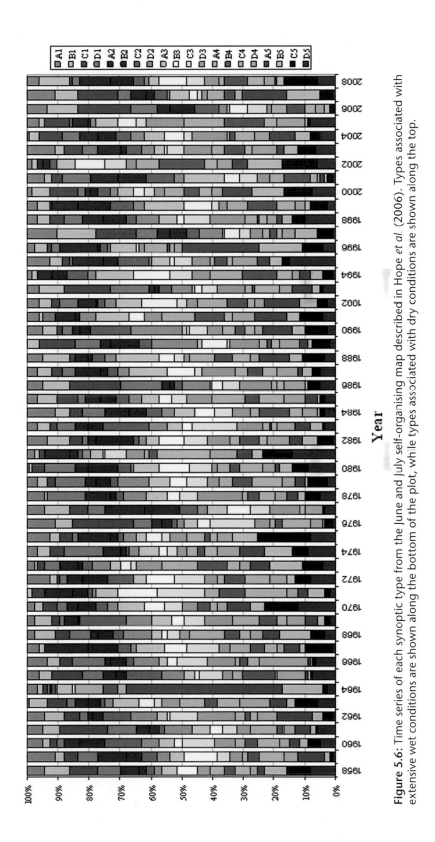

Figure 5.6: Time series of each synoptic type from the June and July self-organising map described in Hope *et al.* (2006). Types associated with extensive wet conditions are shown along the bottom of the plot, while types associated with dry conditions are shown along the top.

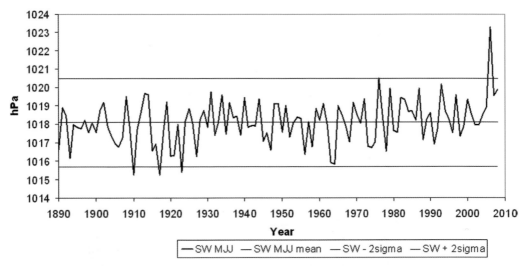

Figure 5.7: Mean sea level pressure (HadSLP2r) averaged over a square over south-west WA for May to July. Also shown are the mean over the 1890 to 2008 period and +/- 2 standard deviations either side of the mean.

for the trend from 1970 to 2008 June–August). This indicates that there has not only been an upward trend in average MSLP across southern Australia, but the frequency of high-pressure systems has also increased markedly.

Comparison with future projections

More than 90% of the coupled climate models submitted for the IPCC Fourth Assessment Report simulate a rainfall decline in the south-west of Australia (IPCC 2007). The decline

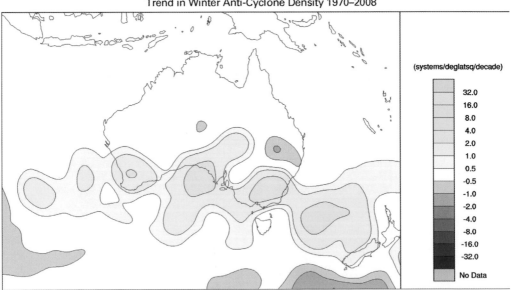

Figure 5.8: Trend in the density of high-pressure systems in June to August from 1970 to 2008. (From Bureau of Meteorology website: www.bom.gov.au/silo/products/cli_chg/)

intensifies with increasing levels of greenhouse gases (CSIRO and Bureau of Meteorology 2007). Using results from the 20th century experiments, Cai and Cowan (2006) determined that 50% of the observed winter rainfall reduction is attributable to anthropogenic forcing. CSIRO and Bureau of Meteorology (2007) show that the recent rainfall declines are of the same order as those expected across the south-west at the end of the century compared to the last 20 years of the 20th century under the A1B scenario (Nakicenovic *et al.* 2000). There is a slight shift in the seasonality of the response, where August has not shown significant declines in the observed record, but does show declines in the projected model climate. The mean climate model response also shows a strong increase in mean sea level pressure (for 2080–2099 compared to 1980–1999) right across the Australian region (Meehl *et al.* 2007). Strong, consistent increases across southern Australia are projected, of the order of 1.5 hPa as an inter-model mean across the south-west. Some models project much greater increases (up to 4 hPa), while others have a more bland response. The observed MSLP record over the south-west shows that the average for the years since 1999 (2000 to 2007) is well above the long-term mean: 1019.2 hPa compared to the long-term average 1890–1999 of 1018.0 hPa, an increase of 1.2 hPa. The average MSLP over the eight-year period from 2000–2007 is also the highest value of any eight-year period within the record. This is due in part to the extreme values in 2006. The frequency of occurrence of synoptic types with high pressure over the region is also projected to increase in four of five climate models assessed by Hope (2006), while the frequency of occurrence of systems with lower pressure over the region are projected to decrease.

These recent observations show that we are well on the way to experiencing what is projected under heightened levels of atmospheric greenhouse gases.

Conclusions

The rainfall decline in south-west Western Australia in the late 1960s was associated with a strong reduction in the number of deep low-pressure systems crossing the region. There was also a contribution from an increase in the frequency of high-pressure systems, but this was of less importance. The recent rainfall declines across the wider south-west region, which, for the far south-west are shown as a step drop across 1999/2000, have been accompanied by a strong increase in the frequency of high-pressure systems. This increase is evident using two distinct methods to define the high-pressure systems. There was almost no change in the number of deep low-pressure systems. The mean sea level pressure in the south-west of Australia has risen, with 2006 being well above anything previously recorded. The spatial extent of the recent rainfall decline is far more widespread than the decline in the late 1960s. This aligns well with the finding that the recent changes were due to an increase in the frequency of high-pressure systems, which impact a wider region of the south-west, rather than a further decrease in the number of low-pressure systems. This difference in how the synoptic systems have altered might suggest an alternative or additional driver of the rainfall change since 2000 compared to the earlier decline.

The recent declines in winter rainfall and increases in winter MSLP are of a similar magnitude to those projected for a mid-range scenario of enhanced levels of atmospheric greenhouse gases at the end of this century.

Acknowledgements

The authors would like to acknowledge the incisive discussion with Bertrand Timbal, and the useful suggestions of three reviewers.

References

Allan R and Ansell TJ (2006) A new globally complete monthly historical gridded mean sea level pressure dataset (HadSLP2): 1850–2004. *Journal of Climate* **19**, 5816–5842.

Cai W and Cowan T (2006) SAM and regional rainfall in IPCC AR4 models: can anthropogenic forcing account for southwest Western Australian winter rainfall reduction? *Geophysical Research Letters*, **33**, L24708, doi:10.1029/2006GL028037.

Cai W, Whetton PH and Karoly DJ (2003) The response of the Antarctic Oscillation to increasing and stabilized atmospheric CO2. *Journal of Climate* **16**, 1525–1538.

CSIRO and Bureau of Meteorology (2007) 'Climate change in Australia – technical report 2007'. CSIRO and Bureau of Meteorology, Australia. <www.climatechangeinaustralia.gov.au>

Frederiksen JS and Frederiksen CS (2007) Inderdecadal changes in southern hemisphere winter storm track modes. *Tellus* **59A**, 599–617.

Frederiksen JS, Frederiksen CS and Osbrough SL (2009a) Modelling of changes in Southern Hemisphere weather systems during the 20th century. In: *18th World IMACS/MODSIM Congress*, 13–17 July 2009, Cairns, Australia, (Eds RS Anderssen, RD Braddock and LTH Newham) pp. 2562–2568. IMACS/MODSIM.

Frederiksen JS, Frederiksen CS, Osbrough SL and Sisson JM (2010) Causes of changing southern hemisphere weather systems. In: *Managing Climate Change: Papers from the GREENHOUSE 2009 Conference.* (Eds I Jubb, P Holper and W Cai) pp. 85–98. CSIRO Publishing, Melbourne.

Fyfe JC (2003) Extratropical Southern Hemisphere cyclones: Harbinger of climate change? *Journal of Climate* **16**, 2802–2805.

Hope P and Foster I (2005) 'How our rainfall has changed – the south-west'. Indian Ocean Climate Initiative, Climate Note 5/05. Indian Ocean Climate Initiative, Perth, Australia.

Hope P, Timbal B and Fawcett R (2009) Associations between rainfall variability in the southwest and southeast of Australia and their evolution through time. *International Journal of Climatology.* doi:10.1002/joc.1627

Hope PK (2006) Projected future changes in synoptic systems influencing southwest Western Australia. *Climate Dynamics* **26**, 765–780.

Hope PK, Drosdowsky W and Nicholls N (2006) Shifts in the synoptic systems influencing southwest Western Australia. *Climate Dynamics* **26**, 751–764.

IOCI (1999) 'Towards understanding climate variability in south western Australia – Research reports on the first phase of the Indian Ocean Climate Initiative'. Technical Report. Indian Ocean Climate Initiative Panel, Perth, Australia.

IOCI (2002) 'Climate variability and change in south west Western Australia'. Technical Report. Indian Ocean Climate Initiative Panel, Perth, Australia.

IPCC (2007) Climate Change 2007: The Physical Science Basis. Contribution of Working Group I to the Fourth Assessment Report of the Intergovernmental Panel on Climate Change. Cambridge University Press, Cambridge, UK.

Jones DA and Weymouth G (1997) 'An Australian monthly rainfall dataset'. *Technical Report No. 70,* Bureau of Meteorology, Australia.

Lavery BM, Joung G and Nicholls N (1997) An extended high-quality historical rainfall dataset for Australia. *Australian Meteorological Magazine* **46**, 27–38.

Li Y, Cai W and Campbell EP (2005) Statistical modeling of extreme rainfall in southwest Western Australia. *Journal of Climate* **18**, 852–863.

Meehl GA, Stocker TF, Collins WD, Friedlingstein P, Gaye AT, Gregory JM, Kitoh A, Knutti R, Murphy JM, Noda A, Raper SCB, Watterson IG, Weaver AJ and Zhao Z-C (2007) 'Global Climate Projections'. In Climate Change 2007: The Physical Science Basis. Contribution of Working Group I to the Fourth Assessment Report of the Intergovernmental Panel on Climate Change. Cambridge University Press, Cambridge, UK.

Nakicenovic N, Alacamo J, Davis G, de Vries B, Fenhann J, Gaffin S, Gregory K, Grubler A, Jung TY, Kram T, La REL, Michaelis L, Mori S, Moita T, Pepper W, Pitcher HM, Price L, Raihi K, Roehrl A, Rogner HH, Sankovski A, Schlesinger M, Shukla P, Smith SJ, Swart R, van Rooijen S, Victor N and

Dadi Z (2000) Emissions scenarios, A special report on Working Group III of the Intergovernmental Panel on Climate Change. Cambridge University Press, UK.

Ryan B and Hope P (Eds) (2005) 'Indian Ocean Climate Initiative Stage 2: Report of Phase 1 Activity. Technical Report'. Indian Ocean Climate Initiative Panel, Perth, Australia.

Ryan B and Hope P (Eds) (2006) 'Indian Ocean Climate Initiative Stage 2: Report of Phase 2 Activity. Technical Report'. Indian Ocean Climate Initiative Panel, Perth, Australia, 37 pp.

Timbal B and Hope P (2008) Observed early winter mean sea level pressure changes over southern Australia: a comparison of existing datasets. *CAWCR Research Letters* **1,** 1–7.

Yin JH (2005) A consistent poleward shift of the storm tracks in simulations of 21st century climate. *Geophysical Research Letters* **32,** 18701.

6. HOW HUMAN-INDUCED AEROSOLS INFLUENCE THE OCEAN–ATMOSPHERE CIRCULATION: A REVIEW

Tim Cowan and Wenju Cai

Abstract

Using targeted ensemble experiments with and without increasing anthropogenic aerosols from coupled ocean–atmosphere climate models, we show that increasing aerosol levels strengthen the globally interconnected ocean current system associated with the Atlantic overturning, inducing a pan-oceanic heat redistribution. This generates a greater poleward shift and intensification of the Agulhas outflow and its retroflection; this process increases the warming rate in the subtropics, and takes heat out of off-equatorial regions generating a cooling. As a result of this poleward shift of the Agulhas, maximum sea surface temperature gradients, mid-latitude storms and the westerly jet shift southward, intensifying the trend of the southern annular mode. These atmospheric circulation responses, in turn, reinforce the ocean circulation changes, constituting an air–sea positive feedback. These results illustrate how northern hemisphere aerosols impact on the southern hemisphere atmospheric circulation trends through a response in ocean circulation.

Introduction

Historically, the impact of volcanic and anthropogenic aerosols has been to partially offset the warming signal due to greenhouse gases. Anthropogenic aerosols, consisting of sulfate, organic carbon, black carbon, nitrate, and dust, have a strong cooling effect on the climate, through both direct (-0.5 Wm^{-2}) and indirect (-0.7 Wm^{-2}) radiative forcings (IPCC 2007), although large uncertainties remain in the magnitude of these forcing components.

Recent observational and modelling studies have provided new insights into the large influence aerosols have on regional-scale climate; for example, Asian aerosols are believed to have contributed to the recent increasing rainfall trend over northern Australia (Rotstayn et al. 2007), although uncertainties remain (Shi et al. 2008). Aerosols, to some extent, have also been blamed for droughts over parts of Africa (Rotstayn and Lohmann 2002), as well as flooding (and droughts) in China during the boreal summertime (Menon et al. 2002). Our understanding of the impacts of aerosols on ocean circulation, however, is less clear. This is due to limitations in our ability to isolate the physical impacts of aerosols on the larger-scale climate; in particular, the paucity of climate models that provide single-purposed experiments allowing an examination of the direct and indirect effect of aerosols.

Previous studies have shown that both volcanic and anthropogenic aerosols have strong impacts on global oceanic heat content and subsequently sea level (Church *et al.* 2005; Delworth *et al.* 2005). There has, however, been little focus shown on the hemispheric scale. In this chapter, we review several recent studies focusing on the following issues (Cai *et al.* 2006; Cai *et al.* 2007; Cai and Cowan 2007): how do aerosols, which are predominantly emitted in the northern hemisphere, impact on ocean circulation changes in the southern hemisphere? Is there a large difference in sea level between the two hemispheres as a result of northern hemisphere anthropogenic aerosols? If there are changes in the ocean, how do these reflect the large-scale changes in atmosphere, and what are the feedbacks responsible? These issues are addressed by comparing the results of coupled ocean-atmosphere climate models using targeted forcing experiments to isolate the aerosol-induced component.

Data, models and experiments

Using targeted experiments from coupled-climate models allows the comparison of climates impacted by a particular forcing (e.g. greenhouse gases, aerosols, ozone depletion). For this purpose, we obtained output from three coupled-climate models that perform these types of experiments over the 20th century. Details of the models used are shown in Table 6.1.

For the CSIRO Mk3 R21 model, an eight-member ensemble with all forcings (greenhouse gases, aerosols, ozone depletion) and a further eight runs with these same forcings except for anthropogenic aerosols (that is, aerosols remain fixed at their 1870 levels) were performed. For the NCAR PCM1 and GFDL CM2.1 models, a set of experiments were performed with only the aerosol component 'turned on' (i.e. greenhouse gases and other forcings were fixed), and the aerosol-induced signal was calculated as the difference between these simulations and a multi-century control run. The respective models' control run was taken out to remove any long-term climate drift due to model spin-up (these models are not flux-corrected). The PCM1 and CM2.1 models do not contain the indirect effects of aerosols, whereas the Mk3 R21 does. In the subsequent sections, we refer to ensemble simulations that contain all forcings as ALL, whereas the simulations that contain all forcings except for anthropogenic aerosols are referred to as AXA. It is the difference between these two sets of ensembles (ALL-AXA) that we define as the aerosol-induced component (Cai *et al.* 2006).

Table 6.1 Models used for analysis of the impacts of aerosols on ocean/atmosphere circulation

Institution	Coupled-climate model	Ocean resolution (lon × lat)	Ensemble members	Reference
Commonwealth Scientific and Industrial Research Organisation (CSIRO)	Mk3 R21	5.6° × 3.2°	8	Cai *et al.* 1997
National Oceanic and Atmospheric Administration/ Geophysical Fluid Dynamics Laboratory (NOAA/GFDL)	CM2.1	1° × 1°	3	Delworth *et al.* 2006
National Center for Atmospheric Research (NCAR)	PCM1	1° × 1°	4	Washington *et al.* 2000

Results

Ocean circulation changes

The inclusion of aerosols in a climate model improves the coherence between models and the observations (Rotstayn *et al.* 2007), which can be seen through the globally averaged changes in air temperature. Figure 6.1 (top panels) shows that including aerosols (blue lines) lowers the global response in air temperature, compared to a scenario where there are no aerosols (orange lines). The AXA simulations in both the GFDL CM2.1 and CSIRO Mk3 R21 overestimate the warming signal particularly from 1950 onwards, when compared to observations (grey line in top panels of Figure 6.1).

Using the aerosol-induced component from the two models' ensemble sets we see that aerosols cause a general decrease in heat fluxes across the northern hemisphere and little change in the southern hemisphere (Figure 6.1, middle panels). As aerosols are predominantly emitted in the northern hemisphere, this trend in the surface heat flux is entirely plausible. However, by studying the aerosol-induced changes in oceanic heat content (Figure 6.1, lower panels) there appears to be very little hemispheric difference for the most part of the 20th century. Aerosols lead to a steady decrease in the oceanic heat content across both hemispheres, regardless of the source of aerosols. One may intuitively expect the oceanic heat content change to follow suit with the surface heat flux change (i.e. small in the southern hemisphere). So what is the basis for this decrease in oceanic heat content across both hemispheres, particularly the cooling in the southern hemisphere, if not via the surface heat fluxes?

Cai *et al.* (2006) studied this problem and discovered that there is an aerosol-induced northward oceanic heat transport across the hemispheres (southern hemisphere to northern hemisphere) via the Global Conveyor Belt. They found, using the CSIRO Mk3 R21 ensemble, that two-thirds of the cross-hemispheric transport takes place in the Atlantic with the rest via the Indo-Pacific region. They also found that the aerosols induce a stronger Atlantic meridional overturning, however recent observations show a slow-down of the deep southward flow of the Atlantic overturning (Bryden *et al.* 2005). Cai *et al.* (2006) emphasised the point that aerosols could be protecting the overturning; that is, without aerosols the slowing would be more pronounced.

The strengthening of the Conveyor can be seen in Figure 6.2, which highlights the trend in the upper 750 m currents (vectors) from the Mk3 R21, and the change in the full-depth vertically averaged temperature (contours) from the three model ensembles (over 1951–2000). Figure 6.2 depicts a strong flow from the south Atlantic to the north Atlantic where the North Atlantic Deep Water formation exists. In this region, the oceanic heat content increases (red colour) as aerosol-induced northward flows converge and contribute to an enhanced local heat loss (to the atmosphere). The northward flow is fed by both a stronger Agulhas outflow (south of Africa) and Indonesian Throughflow. In the Pacific Ocean, a northward flow along the western boundary is evident extending from the south near the Philippines into the central north Pacific, looping until it eventually finds a pathway to the Indian Ocean via the Indonesian Throughflow passage.

The heat transported to the northern hemisphere contributes to the heat loss to the atmosphere seen in Figure 6.1 (middle panels). According to Cai *et al.* (2006), 60% of the heat lost though northern hemisphere oceanic cooling originates from the southern hemisphere oceans. These results suggest aerosols contribute to a pan-oceanic intensification process transporting heat from the southern hemisphere oceans to the northern hemisphere. This also explains why the oceanic heat content in Figure 6.1 (lower panels) and observed sea level rise shows no hemispheric difference.

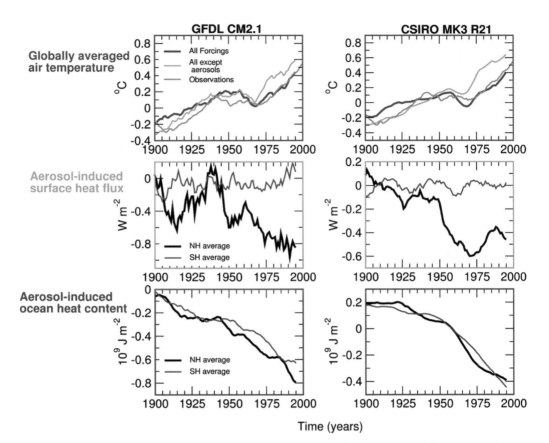

Figure 6.1: Time series over 1900–2000 from the GFDL CM2.1 (left panels) and the CSIRO Mk3 R21 (right panels) of: globally averaged air temperature (top panels) with all forcings (blue line) and all forcings except for aerosols (orange line), aerosol-induced surface heat flux (middle panels) averaged over the southern hemisphere (red line) and northern hemisphere (black line), and the aerosol-induced oceanic heat content (bottom panels) averaged over the southern hemisphere (red line) and northern hemisphere (black line). Observations of globally averaged air temperature from HadCRUT2 (brown line) are shown in the top panels. All data are smoothed using an 11 year running average.

Another important feature aside from the cross-hemispheric heat transport across the equator is the warming south of 40°S. This is due to an aerosol-induced poleward shift in the subtropical gyre, bringing warmer water from the off-equatorial Indian Ocean, enhancing cooling there (Figure 6.2, blue colour). We also see an intensification of the Agulhas outflow and its retroflection (anticyclonic flow). This process has contributed to the observed strong warming of the subtropical Indian Ocean, compared with other subtropical ocean basins (Cai *et al.* 2007).

Atmospheric circulation changes

In response to an aerosol-induced southern hemisphere poleward shift in the oceanic circulation (as a part of the Conveyor intensification) the large-scale atmospheric patterns also move polewards. These changes manifest through the wind stress curl (which maintain the ocean circulation through the Sverdrup balance), and thus through changes in the zonal wind stress. These changes show up as an intensification of the wind and a poleward shift through the

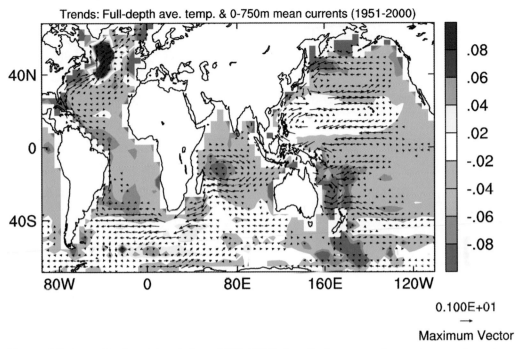

Figure 6.2: Aerosol-induced trends from 1951–2000 of vertically-averaged oceanic currents over 0–750 m (vectors, maximum vector is 1 cm s^{-1} century^{-1} (centimetres per second per century)) from the CSIRO Mk3 R21 and vertically averaged temperature (°C century^{-1}) from the 3-model ensemble.

mid-latitude region of the southern hemisphere, when comparing the ALL scenario with AXA (Figure 6.3). In two separate models in Figure 6.3 (NCAR PCM1 in (a) and (b), and CSIRO Mk3 R21 in (c) and (d)) the inclusion of aerosols (comparing ALL with AXA) leads to a stronger trend in the westerly wind stress, and a slight southward shift. Both trends in ALL and AXA are Southern Annular Mode (SAM)-like, and extend well above the surface reflecting the equivalent-barotropic nature of the trends, as shown in Cai and Cowan (2007).

To understand what drives these atmospheric changes, we first look at the gradients of the models' sea surface temperature (SST) and zonal wind stress trends across the mid- to high-latitude region in the southern hemisphere. This is because any atmospheric poleward shift will be a response to an underlying oceanic shift, which can be determined through gradients of SST. In their study using the CSIRO Mk3 R21 aerosol experiments, Cai and Cowan (2007) showed that the location of the maximum SST and zonal wind stress gradients (Figures 4a and 4c, Cai and Cowan 2007, p. 4) were located further south when aerosols were included. This suggests that mid-latitude storms, including rainfall, also shift poleward as a result of this aerosol impact. We also see from Figure 6.3 (blue colour) that aerosols tend to reduce the westerlies to the north of 40°S, consistent with changes that have been observed in this region (Cai and Cowan 2006). These findings lead us to conclude that aerosols lead to changes in the SAM, but in turn these large-scale atmospheric changes can also feed back into the underlying ocean circulation. As the zonal winds move polewards (and the polar vortex contracts), we expect the Ferrel cell, which describes the integrated effect of the eddy circulation systems (high and low pressure), to also shift southward and strengthen partly as a result of anthropogenic aerosols (Figure 5, Cai and Cowan 2007, p. 4). This Ferrel cell shift affects the SST anomalies, which in turn strengthens this 'Ferrel cell-ocean circulation feedback': a poleward shift and

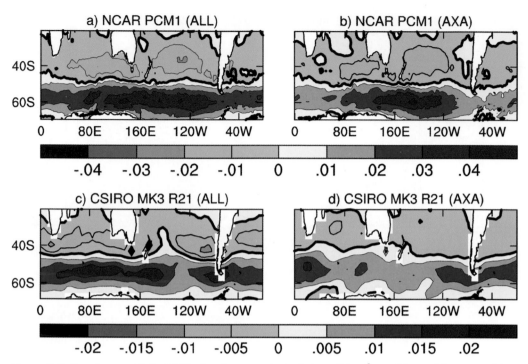

Figure 6.3: Trends over 1941–2000 of surface zonal wind stress (N m^{-2} per 60 years) for ensembles of (a) all forcings (ALL) and (b) all forcings except for aerosols (AXA) for the NCAR PCM1 model; (c) and (d) are the same as (a) and (b), but for the CSIRO Mk3 R21.

intensification of the SAM trend, as shown through changes in the surface zonal wind stress (extending up into the atmosphere), are responsible for a Ferrel cell shift, which in turn feeds back into underlying SSTs, enhancing this feedback. The dynamics of such a feedback, however, need to be thoroughly investigated in a number of models before any reasonable conclusion can be drawn. So far, only the atmospheric component of the aerosol-induced impact on the Ferrel cell has been studied in the CSIRO Mk3 R21. Other forcings such as greenhouse gases and stratospheric ozone depletion are already known to influence climate drivers like the SAM (Cai and Cowan 2006).

Conclusions

In recent years, much work has been undertaken on the impacts of greenhouse gases, stratospheric ozone depletion, volcanic aerosols, and anthropogenic aerosols on the Earth's climate. However, little work has been carried out on examining the impact of aerosols on ocean–atmospheric circulation changes, such as surface warming in the southern subtropical Indian Ocean. Recent targeted simulations from a number of coupled climate model experiments that isolate the impact of anthropogenic aerosols show that increasing aerosols mitigates a greenhouse-induced slowdown of the globally interconnected ocean current system associated with the Atlantic overturning, inducing a cross-hemispheric heat redistribution, cancelling out any hemispheric difference in sea level. This pan-oceanic adjustment includes a greater poleward shift and intensification of the Agulhas outflow and its retroflection; the process increases the warming rate in the subtropics (particularly the southern Indian Ocean) and takes heat out of the off-equatorial region, generating a cooling. Due to this poleward shift, maximum SST gradients, mid-latitude storms and the westerly jet also shift southward, intensifying the trend of the SAM. In turn, the atmosphere reinforces the ocean circulation changes, resulting in the formation of an air–sea positive feedback, through the Ferrel cell. The results illustrate an impact of northern hemisphere aerosols in changing the ocean circulation in the southern hemisphere, which in turn influences the large-scale weather systems. Aerosols are partially responsible for a southward shift of these rain-bearing systems away from southern Australia and other mid-latitude regions, and thus may be one such cause for the persistent drought-like conditions experienced in recent decades across south-east Australia.

References

Bryden HL, Longworth HR and Cunningham SA (2005) Slowing of the Atlantic meridional overturning circulation at 25°N. *Nature* **438**, 655–657.

Cai W, Syktus JI, Gordon HB and O'Farrell SP (1997) Response of a global coupled-atmosphere-sea ice climate model to an imposed North Atlantic high-latitude freshening. *Journal of Climate* **10**, 929–948.

Cai W, Bi D, Church J, Cowan T, Dix M and Rotstayn L (2006) Pan-oceanic response to increasing anthropogenic aerosols: Impacts on the Southern Hemisphere oceanic circulation. *Geophysical Research Letters* **33**, L21707, doi:10.1029/2006GL027513.

Cai W and Cowan T (2006) SAM and regional rainfall in IPCC AR4 models: can anthropogenic forcing account for southwest Western Australian winter rainfall reduction? *Geophysical Research Letters* **33**, L24708, doi:10.1029/2006GL028037.

Cai W, Cowan T, Dix M, Rotstayn L, Ribbe J, Shi G and Wijffels S (2007) Anthropogenic aerosol forcing and the structure of temperature trends in the southern Indian Ocean. *Geophysical Research Letters* **34**, L14611, doi:10.1029/2007GL030380.

Cai W and Cowan T (2007) Impacts of increasing anthropogenic aerosols on the atmospheric circulation trends of the Southern Hemisphere: an air-sea positive feedback. *Geophysical Research Letters* **34**, L23709, doi:10.1029/2007GL031706.

Church JA, White NJ and Arblaster JM (2005) Significant decadalscale impact of volcanic eruptions on sea level and ocean heat content. *Nature* **483**, 74–77, doi:10.1038/nature04237.

Delworth TL, Ramaswamy V and Stenchikov GL (2005) The impact of aerosols on simulated ocean temperature and heat content in the 20th century. *Geophysical Research Letters* **32**, L24709, doi:10.1029/2005GL024457.

Delworth TL, Broccoli AJ, Rosati A, Stouffer RJ, Balaji V, Beesley JA, Cooke WF, Dixon KW, Dunne J, Dunne KA, Durachta JW, Findell KL, Ginoux P, Gnanadesikan A, Gordon CT, Griffies SM, Gudgel R, Harrison MJ, Held IM, Hemler RS, Horowitz LW, Klein SA, Knutson TR, Kushner PJ, Langenhorst AR, Lee HC, Lin SJ, Lu J, Malyshev SL, Milly PCD, Ramaswamy V, Russell J, Schwarzkopf MD, Shevliakova E, Sirutis JJ, Spelman MJ, Stern WF, Winton M, Wittenberg AT, Wyman B, Zeng F and Zhang R (2006) GFDL's CM2 Global Coupled Climate Models. Part I: Formulation and Simulation Characteristics. *Journal of Climate* **19**(5), doi:10.1175/JCLI3629.1.

IPCC (2007) Summary for Policymakers. In: Climate Change 2007: The Physical Science Basis. Contribution of Working Group I to the Fourth Assessment Report of the Intergovernmental Panel on Climate Change. Cambridge University Press, Cambridge, UK.

Menon S, Hansen J, Nazarenko L and Luo Y (2002) Climate effects of black carbon aerosols in China and India. *Science* **297**, 2250–2253.

Rotstayn LD and Lohmann U (2002) Tropical rainfall trends and the indirect aerosol effect. *Journal of Climate* **15**, 2103–2116.

Rotstayn LD, Cai W, Dix MR, Farquhar GD, Feng Y, Ginoux P, Herzog M, Ito A, Penner JE, Roderick ML and Wang M (2007) Have Australian rainfall and cloudiness increased due to the remote effects of Asian anthropogenic aerosols? *Journal of Geophysical Research* **112**, D09202, doi:10.1029/2006JD007712.

Shi G, Cai W, Cowan, T Ribbe J, Rotstayn L and Dix M (2008) Variability and trend of the northwest Western Australia rainfall: observations and coupled climate modeling. *Journal of Climate* **21**, 2938–2959.

Washington WM, Weatherly JW, Meehl GA, Semtner Jr AJ, Bettge TW, Craig AP, Strand Jr WG, Arblaster J, Wayland VB, James R and Zhang Y (2000) Parallel climate model (PCM) control and transient simulations. *Climate Dynamics* **16**, 755–774.

7. FRESHWATER BIODIVERSITY AND CLIMATE CHANGE

Jenny Davis, Sam Lake and Ross Thompson

Abstract

Freshwater ecosystems are particularly vulnerable to the effects of climatic change because many are already degraded by existing stressors. Increasing agricultural production and urban development, fundamentally driven by global population growth, have resulted in major changes in water use and land management. In addition to direct climatic impacts, it is the indirect impacts arising from interactions with existing stressors that may have the largest negative effects on freshwater ecosystems. However, not all scenarios are negative. The 'boom and bust' ecosystems of arid and semi-arid regions (comprising more than 70% of the Australian continent) are highly resilient to climatic variability. For these systems, the protection of refugia, which support populations during dry times, is of utmost importance. Additionally, the processes supporting dispersal and recolonisation also need protection. To reduce the predictive uncertainties created by the confounding effects of multiple pre-existing stressors, long-term monitoring programs are needed at sites where the only major impact is climatic change.

Introduction

The adverse realities of climate change are now recognised (IPCC 2007) and, concomitant with the need to reduce greenhouse gases is the need to develop strategies to mitigate impacts on terrestrial, marine and aquatic ecosystems. Inland waters are particularly vulnerable to climate change because many are already degraded by existing stressors. An estimated 50% loss of wetlands has occurred globally since 1900, as a result of multiple factors including land use change, hydrological change, invasive species and pollution (MEA 2005). The fundamental drivers of this loss are population growth and increasing economic development. Therein large trade-offs exist, as the changes in land use and water use driven by agriculture have also been important in alleviating world hunger and poverty (MEA 2005; Gordon *et al.* 2008). A comparison of potential biodiversity loss by 2100, in marine, terrestrial and freshwater domains, based on scenarios of change in atmospheric carbon dioxide, climate, vegetation and land use, revealed that freshwater ecosystems would experience substantial impacts from land use, biotic exchange and climate (Salo *et al.* 2000). Climate change will create 'top–down' impacts over existing multiple 'bottom–up' impacts to result in further loss of biodiversity and ecosystem change (MEA 2005). Given current impacts and the predicted growth of the global population to 8.1–9.6 billion by 2050 (UNEP 2007) the protection of freshwater biodiversity and ecosystems must be viewed as the ultimate conservation challenge (Dudgeon *et al.* 2006).

Australian landscapes and climate

Morton *et al.* (2009) suggested that the issue of climate change is particularly important for Australia because of the aridity, low soil fertility and high climatic variability that characterise much of the continent. Accordingly, adaptation and mitigation solutions from research undertaken elsewhere, particularly in the northern hemisphere, may not be immediately transferable.

Since the break up of Gondwana, the Australian landmass has undergone massive long-term climatic change. Arid or semi-arid conditions now occur over much (>70%) of the Australian continent (Figure 7.1). Aridity combined with extreme variability in rainfall (often resulting in droughts alternating with floods) has had a great influence on ecological patterns and processes (Orians and Milewski 2007). All but the most permanent of flowing or standing waterbodies can be described as 'boom and bust' systems (Boulton and Brock 1999; Jenkins and Boulton 2003; Kingsford and Norman 2002). These systems are characterised by plants and animals which respond opportunistically to the 'good times' when water is present; germinating and growing, or feeding and breeding, followed by dispersal, aestivation, retreat to refugia or death when dry conditions return. Such ecosystems are highly 'resilient' to climatic variability.

Unfortunately, much of this resilience has already been lost in floodplain rivers in eastern Australia, especially in the Murray–Darling Basin, where rivers are heavily regulated. Flow regimes have been severely compromised to meet water-use demands, especially those of irrigation. The lack of floods has meant the loss of 'boom and bust' ecological dynamics. Innovative approaches are urgently required to ensure that these ecosystems are not completely lost under a drying climate.

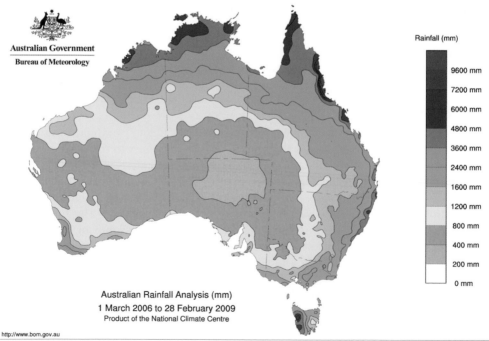

Figure 7.1: Rainfall isohyets reveal that high rainfall in Australia is restricted to northern and eastern coastal regions and the extreme southwest. The remainder of the continent experiences arid or semi-arid conditions. (Source: Bureau of Meteorology)

A major challenge exists to ensure that the pathways for ecosystem recovery following large or episodic rainfall events are maintained in both regulated and unregulated systems. In particular, the protection of refugia, which support populations during dry times, is of utmost importance.

The interaction of climate change and multiple existing stressors

The predicted increase in the frequency and intensity of storms and floods, prolonged periods of drought and increased incidence of wildfires, will create additional stressors on Australian aquatic ecosystems already suffering multiple and chronic impacts. Existing stressors include changes in hydrology, eutrophication, salinisation, acidification, sedimentation, habitat loss and simplification, and the invasion of exotic species (Davis and Froend 1999; Davis and Koop 2006; Lake and Bond 2007; NLWRA 2001).

Predictions that northern Australia will become wetter and southern Australia will become drier suggest that the effects of climate change in the south will be much more severe than in the north (with the possible exception of systems such as Lake Eyre which are filled by distant northern rainfall). Drier conditions will increase the competition for water between human uses and ecosystem requirements. The probable intensity of competition for water suggests that the larger river basins dominated by human activities will be at the greatest risk from climate change (Palmer et al. 2008).

The predicted increase in precipitation in northern Australia has already triggered proposals for new agricultural activities within the region. The perceived economic and social benefits of expanded agricultural and horticultural activity in northern Australia have to be considered against the potentially high environmental impacts, particularly where new impoundments and irrigation schemes on tropical rivers will result in the alteration of existing flow regimes.

The risk of nuisance algal blooms in eutrophic riverine systems will increase under conditions of low flow and higher water temperatures brought about by both water extraction and climate change (Davis and Koop 2006; Viney et al. 2007). Water quality problems associated with decomposing algal blooms include low dissolved oxygen concentrations and algal toxins. These, in turn, can result in fish kills, waterbird deaths, noxious odours and render water unfit for human or stock consumption. However, while flushing flows usually act to improve water quality in riverine ecosystems, nutrients and other pollutants flushed into a standing water-body from the surrounding catchment often result in poorer water quality. For example, water quality monitoring undertaken at eutrophic urban wetlands in south-western Australia has revealed a return to better water quality through a reduction in trophic status associated with a decline in annual rainfall over the last decade. Lower rainfall appears to have resulted in a reduction in the transport of nutrients through the sandy coastal soils comprising the catchments of these wetlands.

The potential effects of a decrease in annual precipitation can be inferred from the effects of current drought conditions in south-eastern Australia and declining annual rainfall in south-western Australia. The latter has been associated with the increasing terrestrialisation of some wetlands. A decrease in the extent and duration of inundation has resulted in the prolific growth of weedy terrestrial species and introduced grasses over large expanses of dry sediments in basin wetlands. Prolonged drought conditions in south-eastern Australia have resulted in the 'lentification' (conversion to still-water ecosystems) of many lowland rivers. River pools often more closely resemble standing waters than flowing waters, with an expansion in the cover of submerged aquatics (*Potamogeton* and *Vallisneria*) and fringing macro-phytes (*Phragmites* and *Typha*).

Perhaps surprisingly, a reduction in annual precipitation appears to have had a beneficial impact on dryland salinisation, a major environmental issue in both south-eastern and south-western Australia. Severe drought in south-eastern Australia appears to have reduced the rate of expansion of dryland salinisation because drier conditions have resulted in declining rather than rising salty water tables (Anderies *et al.* 2006).

Biodiversity impacts – potential 'winners' and 'losers'

Scenarios of increasing temperature and decreased precipitation (IPCC 2007) indicate that alpine species and cool-adapted species (often with Gondwanan affinities) form an important subset of those at risk of extinction in Australian inland waters. Other species, such as those in south-western Australia, may not have the dispersal capabilities to move to more favourable climates in other regions (for example, Tasmania). Hotspots of aquatic biodiversity are likely to be disproportionately affected by climate change due to biogeographic history and limited spatial extent. Increased temperatures will cause a retraction of the range of these species to higher altitudes, which for species already limited to high altitude refugia will result in significant population reductions and the potential extinction of some species. These species will literally have nowhere else to go.

Groups particularly at risk include species of freshwater crayfish, stream-dwelling invertebrates (Blephaceridae and eustheniid stoneflies), the mountain shrimp (*Anaspides tasmaniae*) and a number of obligate temperate fish species. Hotspots of endemism in freshwater fishes, in the highlands of Tasmania (alpine *Galaxias* and *Paragalaxias* species) and in south-western Australia (*Galaxiella* spp., *Lepidogalaxias* spp.) are likely to be particularly vulnerable to increased temperatures. In upland ecosystems there will be little opportunity to mitigate impacts through altered land and water management, whereas some remediation activities may be possible in intensively managed agricultural landscapes.

Climate change 'winners' may result where current negative impacts, usually arising from introduced and invasive species, are reduced by climatic change. Populations of native fishes, such as the galaxiids, which have been reduced by predation and competition from introduced species, may increase as the populations of cool-adapted exotics, such as brown and rainbow trout, decrease. Incipient upper lethal temperatures of 22.5–25°C for brown trout (*Salmo trutta*) and 25–26°C for rainbow trout (*Oncorhynchus mykiss*) (Tilzey 1977) indicate that trout are unlikely to persist on mainland Australia, except in some high altitude areas. This prediction is based on predicted water temperatures and observed range contractions in response to climate change elsewhere (Ficke *et al.* 2007). Opportunities exist to replace trout in warming waters with species such as Macquarie perch (*Macquaria australasica*), trout cod (*Maccullochella macquariensis*) and river blackfish (*Gadopsis marmoratus*). Native species such as silver perch (*Bidyanus bidyanus*), rainbow fish (*Melanotaenia* spp.) and spangled grunter (*Leiotherapon unicolor*) may extend their range southwards.

Warming may promote the expansion of existing invasive species (Dukes and Mooney 1999) which have invaded from warmer and tropical regions. In temperate regions, exotic species such as mosquitofish *(Gambusia holbrooki)* and carp *(Cyprinus carpio)* and, in tropical regions African cichlids, will be favoured. However, prolonged dry conditions can also act to exclude some invasive species. A noteworthy example is the mosquitofish, a species which is often common and abundant in permanent standing or slow-flowing waters throughout temperate Australia. It is a live-bearer (eggs are brooded within the female) with no strategies to survive seasonal drying. This previously abundant species has recently disappeared from wetlands on the Swan Coastal Plain in south-western Australia, which have changed from permanently inundated to seasonally drying systems as annual rainfall has decreased over the last decade. Responses, however, may not be simple. In eastern Australia, mosquitofish have declined to low numbers under drought conditions but populations have recovered rapidly with a return to wetter conditions.

A warmer climate is likely to facilitate a southward invasion of disease-carrying mosquitoes, with increased incidence of Ross River virus, Murray Valley encephalitis and dengue fever (IPCC 2007). Increased mosquito ranges and increased transmission of disease are associated with higher temperatures and the presence of small isolated water bodies in the landscape (Kearney *et al.* 2009; McMichael *et al.* 2006).

Range extensions of warm-water invasive plants including lippia (*Phyla canescens*), arrowhead (*Sagittaria montevidensis*) and water hyacinth (*Eichhornia crassipes*) are likely to have a significant negative impact on inland waters and their riparian zones; however, similar to warm-water fishes, these species thrive in permanent waters, but are often excluded under seasonal and episodic drying regimes. Invasive fringing species already favoured by drier conditions and disturbance, such as the bulrush (*Typha orientalis*), are likely to increase in distribution under drier conditions.

Impacts on coastal ecosystems

Changes in sea level and patterns of rainfall will affect interactions between freshwater and marine environments. Estuaries, coastal lakes and wetlands and mangroves will all be affected by changes in the balance of freshwater and marine inputs. Sea level rise will act to reduce the extent of coastal dune lakes supported by lenses of freshwater overlying saline groundwater. In northern Australia, sea level rise and changes in rainfall have allowed the tidal influence of floodplain rivers to move inland with an increase in bare and saline mud flats and the death of *Melaleuca* forests (Eliot *et al.* 1999). Decreased freshwater inputs to estuaries, especially in droughts, will allow seawater to move further up estuaries into freshwater sections, reducing freshwater biota and infiltrating freshwater groundwater systems.

Ecosystem processes and non-linear effects

Considerable uncertainties exist in predicting the impacts of climate change on freshwater ecosystems. The development and testing of conceptual ecological models that examine the impact of multiple stressors on aquatic ecosystems, and recognise that responses may be non-linear, are now essential activities for identifying critical processes and predicting changes, particularly the possibility of 'ecological surprises' or 'catastrophic regime shifts' (Folke *et al.* 2004; Gordon *et al.* 2008; Scheffer *et al.* 2001; Mayer and Rietkerk 2004).

Generalised versions of the models presented by Gordon *et al.* (2008), which summarised the differences between gradual ecological change and three types of regime shifts based on precipitation–vegetation interactions in agricultural zones, provide a useful starting point for examining potential climate-driven changes (Figure 7.2). Gordon *et al.* (2008) suggested that agricultural modification of water flows produced a variety of ecological regime shifts that operate across a range of spatial and temporal scales. Ecological dynamics were defined both

This approach is already being adopted through the National Climate Change and Adaptation Research Facility supported by the Commonwealth Department of Climate Change.

2. Utilisation of long-term datasets.

Pre-existing long-term datasets currently collected for other purposes (eg. water quality assessment) need to be interrogated to determine trends arising from climatic impacts.

3. Establishment of long-term and large-scale monitoring programs and 'sentinel' sites.

Given the confounding effect of multiple pre-existing stressors, the predictive uncertainties associated with the impacts of climate change on aquatic ecosystems could be reduced by monitoring ecosystem components and processes at a series of sites where the only major impact will be climatic. The ideal sites for the collection of long-term data are those where current programs collecting climatic, hydrological and ecological datasets are already underway (Figure 7.3). Likens *et al.* (2009) advocated the establishment of long-term and large-scale programs to understand aquatic ecosystem structure and processes across the full range of hydrological and climatic variability that characterises the Australian continent.

Possible sentinel sites include the Franklin River and Central Plateau Lakes in the Tasmanian World Heritage Wilderness Area, relict streams within the George Gill and West MacDonnell Ranges in central Australia, selected rivers and wetlands within protected areas in northern Australia (the Kimberley, Kakadu, Arnhem Land and Cape York regions), headwater streams in protected areas along the Great Divide, dune lakes and estuarine-freshwater systems within protected coastal areas. Although a program to track climate-driven ecological change in inland waters will require new funding, the ultimate economic costs associated with not truly distinguishing climatic impacts from other stressors, will be far greater.

4. Establishment of a climate change observation network for inland waters.

The trends identified from pre-existing sites and new sentinel sites would form a climate change observation network for inland waters and should be held within a publicly accessible database. This approach expands the recommendation of Likens *et al.* (2009) for the establishment of national archives for long-term data and samples and clear custodial arrangements to protect, update and facilitate knowledge of aquatic ecosystems.

5. Development of downscaled climate change models that will facilitate local and regional responses.

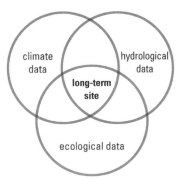

Figure 7.3: Ideal long-term sites are those where programs for the collection of climatic, hydrological and ecological data intersect.

Aquatic ecologists need to work closely with climatologists and hydrologists to ensure that realistic scenarios form the basis of well-coordinated local and regional mitigation activities.

6.   Development of strategies to maximise the resilience of freshwater ecosystems.

The importance of maximising resilience, which includes identifying and protecting refugia, cannot be overstated. Refugia in many systems, especially regulated rivers and streams, are compromised because migration to and from refugial pools is prevented by numerous barriers. These include dams, weirs and levees to and from floodplains. This loss of connectivity, both longitudinal and lateral, in many waterways will serve to exacerbate the effects of climate-induced drying.

The following strategies will all contribute to maximising resilience:

- development of conceptual ecological models to identify non-linear dynamics and critical thresholds;
- reduction of existing stressors through on-ground activities and the utilisation of existing policy tools and initiatives;
- implementation of a strategic program of large-scale habitat restoration to facilitate biotic redistributions, and
- implementation of targeted species restoration programs (and translocation programs where appropriate and feasible).

Conclusions

There is an urgent need to recognise and reduce the impacts of pre-existing stressors on Australian inland waters to mitigate the risks of climate-induced changes to vulnerable freshwater species and ecosystems. This was recently emphasised by Chessman (2009) who identified the absence of stream shading, through the historical loss of riparian vegetation, and reductions in stream flow created by water withdrawals, as major compounding stressors in New South Wales rivers. These stressors, which affect almost all riverine ecosystems throughout temperate Australia, indicate that riparian revegetation programs and increased environmental water allocations must be an important part of climate change mitigation strategies throughout southern Australia. In addition, strategic planning and development of policies to protect in-stream flow regimes in the advent of agricultural expansion, are likely to be of major importance for northern inland waters.

The new paradigm for land and water management in Australia, proposed by Likens *et al.* (2009), in which science and management moves from a reactionary approach to strategic problem solving, with sufficient knowledge to avoid problems, or address them before they arise, must be supported. This is vital if we are to avoid a catastrophic loss of aquatic ecosystem components, processes and services as the direct and indirect consequences of climate change impacts.

Acknowledgements

The support of colleagues and facilities at Monash University and Murdoch University, on projects and in discussions which contributed to the thoughts expressed in this paper, are gratefully acknowledged.

References

Anderies JM, Ryan P and Walker BH (2006) Loss of resilience, crisis and institutional change: lessons from an intensive agricultural system in southeastern Australia. *Ecosystems* **9**, 865–878.

Boulton AJ and Brock MA (1999) *Australian Freshwater Ecology – Processes and Management.* Gleaneagles Publishing, Glen Osmond, Australia.

Chessman BC (2009) Climate changes and 13 year trends in stream macroinvertebrate assemblages in New South Wales, Australia. *Global Change Biology,* doi:10.1111/j.1365-2486-2008.01840.x

Davis JA and Froend R (1999) Loss and degradation of wetlands in southwestern Australia: underlying causes, consequences and solutions. *Wetlands Ecology and Management* **7**, 13–23.

Davis JR and Koop K (2006) Eutrophication in Australian rivers, reservoirs and estuaries – a southern hemisphere perspective on the science and its implications. *Hydrobiologia* **559**, 23–76.

Dudgeon D, Arthington AH, Gessner MO, Kawabata ZI, Knowler DJ, Leveque C, Naiman RJ, Prieur-Richard AH, Soto D, Stiassny ML and Sullivan CA (2006) Freshwater biodiversity: importance, threats, status and conservation challenges. *Biological Reviews* **8**, 163–182.

Duigan C, Orr H, Sime I, McCall R and Clarke S (2007) Climate change impacts on freshwater ecosystems in the UK: research priorities to support adaptation responses. In: *Climate Change and Aquatic Ecosystems in the UK: Science Policy and Management.* (Eds M Kernan, RW Battarbee and HA Binney) pp. 42–43. ENSIS Publishing, University College, London.

Dukes JS and Mooney HA (1999) Does global change increase success of biological invaders? *Trends in Ecology & Evolution* **14**, 135–139.

Eliot I, Finlayson CM and Waterman P (1999) Predicted climate change, sea-level rise and wetland management in the Australian wet-dry tropics. *Wetlands Ecology and Management* **7**, 63–81.

Ficke AD, Myrick CA and Hansen LJ (2007) Potential impacts of global climate change on freshwater fisheries. *Reviews in Fish Biology and Fisheries* **17**, 581–613.

Folke C, Carpenter S, Walker B, Scheffer M, Elmqvist T, Gunderson L and Holling CS (2004) Regime shifts, resilience and biodiversity in ecosystem management. *Annual Review of Ecology, Evolution and Systematics* **35**, 557–581.

Gordon LJ, Peterson GD and Bennett EM (2008) Agricultural modifications of hydrological flows create ecological surprises. *Trends in Ecology and Evolution* **23**, 211–219.

Heath L (2009) Climate factors that lead to fish kills in Australian coastal acid sulfate soil catchments. IOP Conference Series – *Earth and Environmental Science* **6**, doi:10.1088/1755-1307.6/5/352022

IPCC (2007) Intergovernmental Panel on Climate Change: fourth assessment report (AR4). <www.ipcc.ch>

Jenkins KM and Boulton AJ (2003) Detecting impacts and setting restoration targets in arid zone rivers: aquatic micro-invertebrate responses to reduced floodplain inundation. *Journal of Applied Ecology* **44**, 823–832.

Kearney M, Porter WP, Williams C, Ritchie S and Hoffmann A (2009) Integrating biophysical models and evolutionary theory to predict climatic impacts on species' ranges: the dengue mosquito *Aedes aegypti* in Australia. *Functional Ecology* **23**, 528–538.

Kingsford RT and Norman FI (2002) Australian waterbirds – products of the continent's ecology. *Emu* **102**, 47–69.

Lake PS and Bond N (2007) Australian futures: freshwater ecosystems and human water usage. *Futures* **39**, 288–305.

Likens GE, Walker KF, Davies PE, Brookes J, Olley J, Young WJ, Thoms MC, Lake PS, Gawne B, Davis JA, Arthington AH, Thompson R and Oliver RL (2009) Ecosystem science: toward a new paradigm for managing Australia's inland aquatic ecosystems. *Marine and Freshwater Research* **60**, 271–279.

Maher K and Davis JA (2009) 'Ecological Character Description for the Thomsons and Forrestdale Lakes Ramsar Sites'. Report to the Department of Environment and Conservation, Western Australia.

Mayer AL and Rietkerk M (2004) The dynamic regime concept for ecosystem management and restoration. *Bioscience* **54**, 1013–1020.

McMichael AJ, Woodruff R and Hales S (2006) Climate change and human health: Present and future risks. *Lancet* **367**, 859–869.

Millenium Ecosystem Assessment (2005) *Ecosystems and Human Wellbeing: Wetlands and Water.* World Resources Institute, Washington, DC. <www. milleniumassessment.org>.

Morton SR, Hoegh-Guldberg O, Lindenmayer, DB, Harriss Olson M, Hughes L, McCulloch, MT, McIntyre S, Nix HA, Prober SM, Saunders DA, Andersen AN, Burgman MA, Lefroy EC, Lonsdale WM, Lowe I, McMichael AJ, Parslow JS, Steffen W, Williams JE and Woinarski JCZ (2009) The big ecological questions inhibiting effective environmental management in Australia. *Austral Ecology* **34**, 1–9.

Moss B (1990) Engineering and biological approaches to the restoration from eutrophication of shallow lakes in which aquatic plant communities are important components. *Hydrobiologia* **200/201**, 67–378.

National Land and Water Resources Audit (2001) Water Resources in Australia.

Orians GH and Milewski AV (2007) Ecology of Australia: the effects of nutrient-poor soils and intense fires. *Biological Reviews* **82**, 393–423.

Palmer MA, Reidy Liermann CA, Nilsson C, Flörke M, Alcamo J, Lake PS and Bond N (2008) Climate change and the world's river basins: anticipating management options. *Frontiers in Ecology and the Environment* **6**, 81–89.

Sala OE, Chapin III SF, Armesto JJ, Berlow E, Bloomfield J, Dirzo R, Huber-Sanwald, E, Huenneke LF, Jackson RB, Kinzig A, Leemans R, Lodge DM, Mooney HA, Oesterheld M, Poff NL, Sykes MT, Walker BH, Walker M and Wall DH (2000) Global Biodiversity Scenarios for the Year 2100. *Science* **287**, 1770–1774. doi:10.1126/science.287.5459.1770

Scheffer M (1989) Alternative stable states in eutrophic, shallow freshwater systems: a minimal model. *Hydrological Bulletin* **23**, 73–83.

Scheffer M (1998) *Ecology of Shallow Lakes.* Chapman and Hall, London.

Scheffer M and Carpenter SR (2003) Catastrophic regime shifts in ecosystems: linking theory to observation. *Trends in Ecology and Evolution* **18**, 648–656.

Scheffer MS, Carpenter J, Foley C, Folkes B and Walker B (2001) Catastrophic shifts in ecosystems. *Nature* **413**, 591–596.

Schröder A, Persson L and De Roos AM (2005) Direct experimental evidence for alternative stable states: a review. *Oikos* **110**, 3–19.

Tilzey RD (1977) Key factors in the establishment and success of trout in Australia. *Proceedings of the Ecological Society of Australia* **10**, 97–105.

United Nations Environment Programme (2007) Global Environmental Outlook – environment for development (GEO-4). <www.unep.org/geo/geo4/>

Viney NR, Bates BC, Charles SP, Webster IT and Bormans M (2007) Modelling adaptive management strategies for coping with the impacts of climate variability and change on riverine algal blooms. *Global Change Biology* **13**, 2453–2465.

8. CAUSES OF CHANGING SOUTHERN HEMISPHERE WEATHER SYSTEMS

Jorgen Frederiksen, Carsten Frederiksen, Stacey Osbrough and Janice Sisson

Abstract

The changes in southern hemisphere autumn and winter storms have been studied using reanalysed data for May and July climates during the 20th century. The main mechanism of storm formation is related to the vertical shear in the atmospheric winds and is commonly known as baroclinic instability. For July, storm track modes growing on the subtropical jet show a dramatic reduction in growth rate post-1975. This reduction in the intensity of storm development has continued to the present time for storms that cross Australia, and is associated with the observed decrease in rainfall in southern Australia. For May, the strength of the subtropical storm track crossing Australia has decreased while the polar storm track has increased. This again is associated with a decrease in rainfall across southern Australia. These effects have become more pronounced with time. We find for both autumn and winter that the rainfall reduction is also associated with a decrease in the vertical mean latitudinal temperature gradient and in the peak upper tropospheric jet stream zonal winds near 30° south throughout most of the southern hemisphere (SII).

We also examine the ability of climate models to simulate the observed reduction in the vertical shear, or baroclinic instability, of the SH subtropical jet. The Phillips (1954) criterion is a useful diagnostic of incipient baroclinic instability and we have considered this diagnostic in four cases; one involving the changes in the latter half of the 20th century, and three involving changes between the end of the 20th century and different base periods in pre-industrial simulations. The climate models display quite disparate abilities to simulate this diagnostic. Projected changes in baroclinic instability from skilful models suggest that further large reductions are possible under different greenhouse gas emission scenarios, especially over the Australian region. By implication, this suggests further reductions in storm development and further reductions in rainfall, over southern Australia.

Introduction

During the second half of the 20th century, there have been significant reductions in the rainfall across southern Australia and particularly across south-west Western Australia (SWWA) (Hope *et al.* 2006; Nicholls 2007; Bates *et al.* 2008; Cai and Cowan 2008; Ummenhofer *et al.* 2008). Frederiksen and Frederiksen (2005, 2007) suggested that the decrease in winter rainfall was largely due to a reduction in the intensity of storm development over southern Australia towards the end of the 20th century. The main mechanism for storm development is related to the vertical wind shear in the region of the subtropical and polar jet streams. Frederiksen and Frederiksen (2005, 2007) showed that a reduction in the vertical wind shear, associated with changes in the jet stream strength, resulted in a dramatic reduction in the growth rates of winter storms crossing southern Australia. Frederiksen (2007) reviews the literature on storm formation.

In this chapter, the changes in southern hemisphere autumn and winter storm track modes during the 20th century have been studied based on reanalysed observations and on data from climate models. We examine whether the reduction in the growth rate of winter storms has continued into the present time. We also study the corresponding changes in autumn, focusing on the month of May. In addition, we evaluate the response of the Coupled Model Intercomparison Project Three (CMIP3) climate models (Meehl *et al.* 2007) to observed natural and anthropogenic forcing, including increasing greenhouse gases, from pre-industrial to the end of the 20th century. The extent to which the models show similar atmospheric winter circulation changes, as seen in Frederiksen and Frederiksen (2007) for the reanalysis data, is discussed, as well as the implications of these results for climate change projections and attribution studies.

Changes in the southern hemisphere circulation

The southern hemisphere July climates for the periods 1949–1968 and 1975–1994, using the National Centers for Environmental Prediction (NCEP) reanalyses, have been compared and there are significant differences between the two periods. Most noticeable is a reduction of about 17% in the peak strength of the SH subtropical jet stream (Figure 1, Frederiksen and Frederiksen 2007, p. 602). Similar changes occur in May between 1949–1968 and the later periods 1975–1994 and 1997–2006. In May, however, the decrease in the strength of the subtropical jet (near 30°S) is slightly less than in July; conversely, the increase, in May, in the polar jet (near 60°S) is slightly more than in July (not shown). In both periods, there is a maximum in the zonal wind strength in the subtropics (near 30°S) at about the 200 hPa pressure level.

The wind changes are directly associated with changes in the Hadley circulation in the southern hemisphere. The thermal structure of the SH atmosphere has also changed with a significant warming south of 30°S, tending to reduce the equator–pole temperature gradient (not shown). Such changes would be expected to have a significant effect on the stability of the SH circulation and hence on the nature of the SH storms, which have a major impact on southern Australia, and other modes of weather variability. In fact, in both July (Figure 8.5 below) and May (not shown), the SH atmosphere has generally become less unstable in those regions associated with the generation of mid-latitude storms.

Changes in storm track modes from reanalyses

An analysis of the impact of these observed SH winter climate changes on the nature of the dominant SH weather modes was conducted, with particular emphasis on the storm track modes. Here we have used the primitive equation instability model described in Frederiksen and Frederiksen (2005, 2007) to identify the dominant unstable weather modes in each period. In the earlier period, the fastest growing weather mode is a SH storm mode which affects southern Australia, and has largest impact over SWWA (Figure 8.1). This mode consists of a series of eastward propagating troughs (blue shading) and ridges (red shading), and is shown in the top panel of Figure 8.1 at a particular instance. As the troughs and ridges move eastward they amplify to reach a maximum in preferred regions.

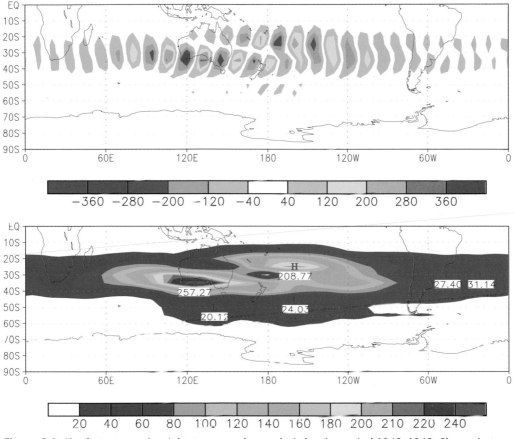

Figure 8.1: The fastest growing July storm mode, mode 1, for the period 1949–1968. Shown in top panel are the 300 hPa troughs (blue) and ridges (red) of the stream function at a particular instance and in bottom panel the corresponding amplitude for the storm track. Units are relative.

The bottom panel of Figure 8.1 shows the storm track associated with this mode and indicates that its largest impact (red shading) is over south-western Australia. By contrast, in the latter period between 1975 and 1994, the dominant SH storm mode has a different horizontal structure. In particular, this weather mode effectively bypasses SWWA and has maximum impact in the central south Pacific (not shown). There are, however, other subdominant weather modes (mode 9 in Table 8.1), with a similar structure and frequency to the dominant mode from the earlier period, but their growth rates have been reduced 33% or more, as shown in Table 8.1. This is consistent with the observed reduction in rainfall over southern Australia, and in particular, SWWA. Figure 8.2 shows the corresponding storm track mode (mode 12) for the recent period 1997–2006. We note from Table 8.1 that compared with the results for 1949–1968, there has been a further reduction in the growth rate (by 37%). The reductions in growth rates of the leading storm track modes in later periods, described above, also occurs for averages taken over the leading 10 or 20 modes that cross Australia, indicating that it is a robust result.

For May we have calculated the average root mean square (RMS) amplitude of the 20 fastest growing storm track modes (with each mode normalised to having the same RMS amplitude) for each of the three periods 1949–1968, 1975–1994 and 1997–2006. The reduction in the

Table 8.1   Properties of leading storm track modes crossing southern Australia

Basic state	Mode	Correlation with mode 1	Growth rate, ω_i	Change in ω_i
1949–1968 NCEP	1	1.0000	0.423 day^{-1}	0.0%
1975–1994 NCEP	9	0.9148	0.282 day^{-1}	–33.5%
1997–2006 NCEP	12	0.8960	0.266 day^{-1}	–37.1%

average growth rate of the 20 fastest growing storm track modes is only about 10% in the latter periods. However, there are dramatic changes in their structures. The results are shown in Figure 8.3. We note that for the early period 1949–1968, the principal storm track is at the latitudes of the subtropical jet; it crosses Australia and has its maximum downstream. In contrast, for 1975–1994, the subtropical storm track is reduced in amplitude and the polar storm track is strengthened. This reduction of the strength of the storm track crossing Australia continues in the period 1997–2006. The RMS associated divergence field across Australia is also reduced correspondingly in the latter periods, as shown in Figure 8.4. Because the divergence is directly proportional to rainfall, this explains the observed reduction in rainfall over southern Australia that has occurred in the latter part of the 20th century.

Changes in baroclinic instability

The Phillips (1954) criterion, generalised for spherical geometry, is a simple diagnostic that provides a measure of incipient baroclinic instability and can be used to identify geographical regions of possible storm development (Frederiksen, 1979, Frederiksen and Frederiksen, 1992). This criterion may be written as

$$(\overline{u}^{(300)} - \overline{u}^{(700)}) - u_c^s \geq 0$$

Here, $\overline{u}^{(300)}$ and $\overline{u}^{(700)}$ represent the 300 hPa and 700 hPa zonal wind, and u_c^s the critical vertical wind shear for instability, that depends on latitude (see Eqn. (3.3) in Frederiksen 1979, p. 197, or Eqn. (3.1) in Frederiksen and Frederiksen 1992, p. 1448), for a given climate. Near the equator, the left hand side of the equation that appears in this chapter is always negative and therefore this criterion is mostly relevant for the development of extra-tropical storms.

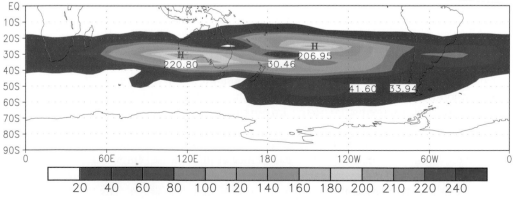

Figure 8.2: As in bottom panel of Figure 8.1 for July mode 12 for 1997–2006.

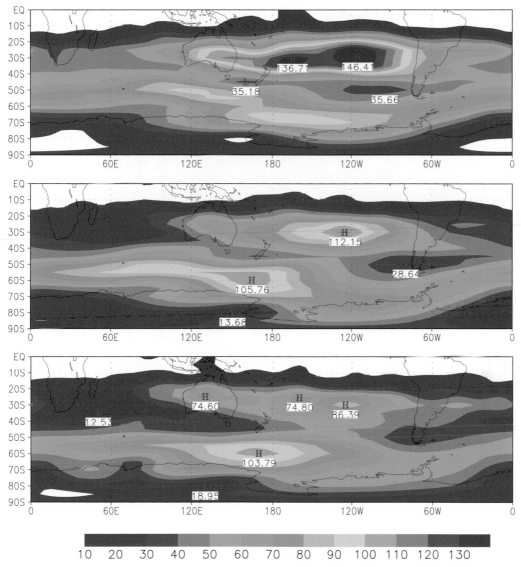

Figure 8.3: RMS 300 hPa stream function averaged over 20 fastest growing May storm track modes for 1949–1968 (top panel), 1975–1994 (middle panel) and 1997–2006 (bottom panel).

In Figure 8.5, we show, for the NCEP reanalysis, regions where this baroclinic instability criterion is positive for the 1949–1968 and 1975–1994 climates, and their difference respectively. For both July climates, these regions coincide with the subtropical jet and a maximum in the criterion occurs in the South Pacific near 30°S. The difference plot (c) shows a reduction in the criterion in the latter period that extends across the whole hemisphere in a band centered near 30°S. As discussed in Frederiksen and Frederiksen (2007), this is associated with a reduction of storm development throughout this band and the reduction in growth rate of the SH climatological storm track modes. Essentially, the SH has become less baroclinically unstable near 30°S in the latter period compared with the former. In contrast, poleward of about 45°S there is an increase in baroclinic instability, especially south and upstream of Australia. Importantly, there is a reduction of about 4.9 ms^{-1} situated over SWWA.

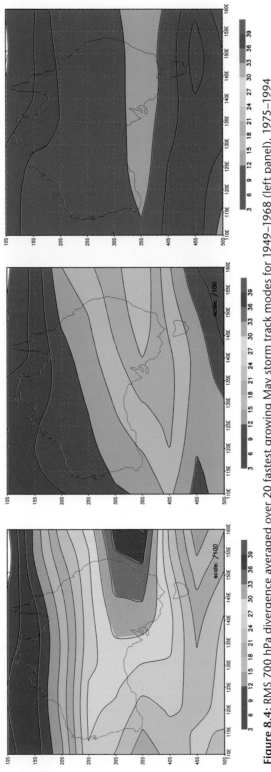

Figure 8.4: RMS 700 hPa divergence averaged over 20 fastest growing May storm track modes for 1949–1968 (left panel), 1975–1994 (middle panel) and 1997–2006 (right panel).

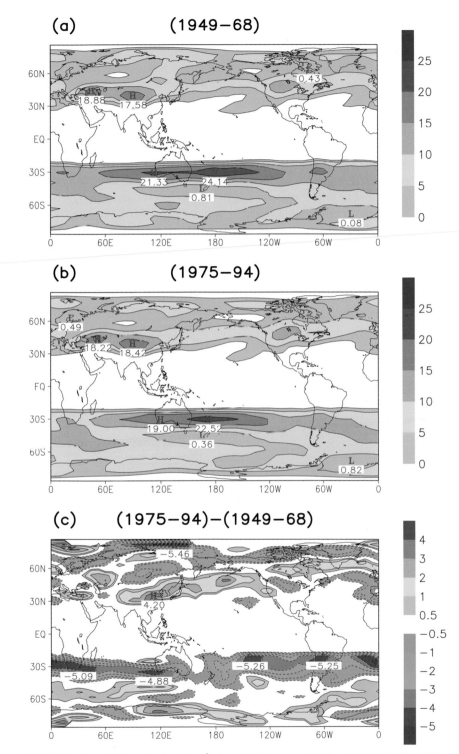

Figure 8.5: Phillips instability criterion (ms^{-1}) from NCEP reanalysis for July (a) 1949–1968; (b) 1975–1994, and (c) the difference (b)-(a).

In Figure 8.6, we show for 22 of the CMIP3 models (see Randall *et al.* 2007 and Meehl *et al.* 2007, for model nomenclature and description) the anomaly pattern correlation (APC), calculated over the domain (60°E–150°E, 45°S–15°S) between the NCEP difference in the Phillips criterion (Figure 8.5(c)) and similar differences calculated with model data in four different ways. The black bars in Figure 8.6 are the APCs with differences in the Phillips criterion calculated for models using the same two 20-year periods (1949–1968 and 1975–1994) as for the NCEP reanalysis (i.e. the 20C3M simulations, Meehl *et al.* 2007). However, because the timing of simulated changes in coupled models may not necessarily synchronize with the reanalysed observations, we have also included APCs with model differences between pre-industrial control runs (i.e. the PICNTRL simulations, Meehl *et al.* 2007) and the (1980–1999) period of the 20C3M runs. This will give an indication of the impact of all the 20th century greenhouse gas forcing. Also, we are interested in the sensitivity of our results to the base period chosen in the PICNTRL runs, and the possible influence of decadal variability on our results. For this reason, we have used three adjoining 20-year periods at the end of the PICNTRL runs, separated by 20 years. These are designated PICNTRL (green bar), PICNTRL_20 (blue bar) and PICNTRL_40 (red bar).

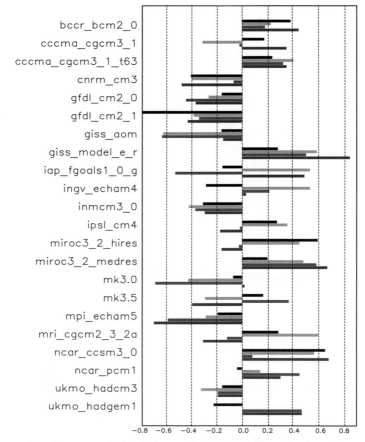

Figure 8.6: Anomaly pattern correlation between the NCEP Phillips criterion difference (Figure 8.5 c) and model Phillips criterion difference for (i) (1975–1994) – (1949–1968) (black bar); (ii) (1980–1999) – PICNTRL (green bar); (iii) (1980–1999) – PICNTRL_20 (blue bar), and (iv) (1980–1999) – PICNTRL40 (red bar).

There is considerable variability in the models' ability to simulate the reanalysis results. About a third of the models show a consistently negative APC in all four cases (e.g. *gfdl_cm2_1*, *giss_aom*, *mpi_echam5*, etc.). For these models, there is an increase in baroclinic instability that would lead to an increase in growth of the storm track modes; however, about a third of the models show a consistently positive APC (e.g. *miroc3_2_medres*, *giss_model_e_r*, *ncar_ccsm3_0*, etc.). There is evidence of a component of decadal variability in the model results.

For the remaining part of this chapter, we are going to concentrate on the results from the *miroc3_2_medres* model. This model agrees fairly consistently with the NCEP reanalysis changes seen in both diagnostics. For example, Figure 8.7 shows the Phillips criterion differences for this model in all four cases. The difference between the (1980–1999) period and the last 20 years of the PICNTRL run is remarkably similar to the NCEP changes (Figure 8.5(c)) throughout the subtropical SH but of somewhat smaller magnitude.

Future projected changes in baroclinic instability

In this section we discuss projections of possible changes in baroclinic instability under three different climate change scenarios using the *miroc3_2_medres* model. The three scenarios we will consider are the Special Report on Emission Scenarios (SRES) B1, A1B and A2 (see, for example, Meehl *et al.* 2007). These involve low, medium and high CO_2 concentrations of 550 ppm, 700 ppm and 820 ppm, respectively, by 2100. Figure 8.8 shows the changes in Phillips criterion for (1980–1999) – PICNTRL, SRESB1 – PICNTRL, SRESA1B – PICNTRL and SRESA2 – PICNTRL, using this model. The period (2080–2099) has been used for each of the SRES scenarios. As the CO_2 concentrations increase, there are progressively larger reductions in the subtropical baroclinic instability, especially over the Australian region. In the SRESA1B and SRESA2 scenarios, these differences are about twice those seen in the model for the 20th century run. This suggests further reductions in the growth rate of the SH storm track modes and a worsening of drought conditions over southern Australia. We have also looked at projected changes in other models that show good correspondence with the reanalysis results. For example, Figure 8.9 shows changes in the Phillips criterion for SRESA2 – PICNTRL for the *giss_model_e_r* and *ncar_ccsm3_0* models. These results are qualitatively similar to those from the *miroc3_2_medres* model, with large reductions over southern Australia.

Discussion and conclusions

There has been approximately a 20% reduction in autumn and winter SWWA rainfall. A 17% reduction in peak July jetstream strength has also been observed. Similar changes in the circulation have occurred in May. In July there has been about 30% reduction in growth rates of leading SH storm track modes crossing Australia during the second half of the 20th century. These changes have continued and spread to south-east Australia during the period 1997–2006. The structures of leading July storm track modes crossing Australia are very similar during the 20th century. A primary cause of the rainfall reduction over SWWA since 1975 is the reduction in the intensity of storm formation and the southward deflection of some storms. In May the strength of the subtropical storm track crossing Australia has decreased during the second half of the 20th century while the polar storm track has increased. Again, these changes in the May storm track modes are consistent with the observed drying across southern Australia.

Meehl GA *et al.* (2007) The WCRP CMIP3 multimodel dataset: a new era in climate change research, *Bulletin of the American Meteorological Society* **88**, 1383–1394, doi:10.1175/BAMS-88-9-1383.

Nicholls N (2007) *Detecting, Understanding and Attributing Climate Change.* Australian Greenhouse Office Publication.

Phillips NA (1954) Energy transformations and meridional circulations associated with simple baroclinic waves in a two-level, quasi-geostrophic model. *Tellus* **6**, 273–286.

Randall DA *et al.* (2007) Climate models and their evaluation. In: Climate Change 2007: The Physical Science Basis. Contribution of Working Group I to the Fourth Assessment Report of the Intergovernmental Panel on Climate Change. (Ed. S Solomon *et al.*), pp. 589–662. Cambridge University Press, Cambridge, UK.

Ummenhofer CC, Sen Gupta C, Pook MJ *et al.* (2008) Anomalous rainfall over southwest Western Australia forced by Indian Ocean sea surface temperatures. *Journal of Climate* **21**, 5113–5134.

Part 2

Impacts and adaptation

9. AUSTRALIAN AGRICULTURE IN A CLIMATE OF CHANGE

Mark Howden, Steven Crimp and Rohan Nelson

Abstract

Australian agricultural systems are likely to be significantly affected by current and future climate changes. These changes will occur in the context of a variety of other issues, requiring adaptive management and policy responses that reduce vulnerabilities to a range of factors. We demonstrate how three such factors (climate change, population growth and consumption patterns) could interact to affect wheat exports both negatively and significantly, and how adaptation could reduce this risk. We discuss how there are many farm-level adaptations to existing systems that could bring substantial benefit in the early stages of climate change. However, these benefits tend to plateau with larger degrees of climate change (approximately above 2°C), requiring more transformational changes to agriculture. These transformations may include change in land use, change in location of agricultural activities or increased diversification of income streams. The role of science in adapting Australian agriculture to climate change is most likely to be enhanced by effective partnerships and engagement processes with farmer groups and other on-ground decision makers who can identify barriers to adoption and what may be maladaptive practices. There is also a strong case for increasing engagement with industry and government policy groups to integrate climate change adaptation with other concerns including mitigation, and also in the exploration of alternative governance arrangements.

Introduction: climate change in the context of global changes

The evolution of agricultural systems in Australia has been shaped by both long-term climatic changes and year-to-year climate variability. This is evident in the historical and ongoing choice of enterprise mix by more than 140 000 farmers that determine land use and the location of specific agricultural industries. Climate interacts with factors such as soil types, management, technology, input costs and product prices to influence production levels, product quality and profitability. Consequently, if Australia experiences a significant shift in climate, systemic changes in Australian agriculture will occur.

This chapter provides an overview of farm-level adaptation responses that may be appropriate for Australian agriculture in response to a broader set of impacts. In particular, we explore both incremental adaptations and more transformative changes and the relationships between these. We do not detail the broader development of adaptive capacity in agricultural communities (see, for example, Nelson *et al.* 2007, 2009), nor the policy responses and governance structures that may support this (see, for example, Nelson *et al.* 2008) due to space

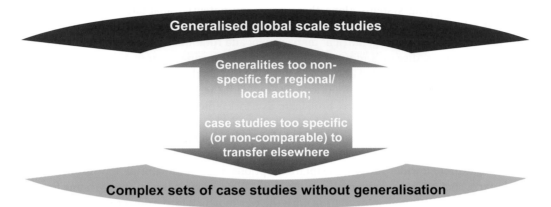

Figure 9.3: The need to seek necessary but sufficient complexity to inform decisions between generalised, large-scale studies with little local relevance and complex sets of local studies that are difficult to generalise.

These types of intensive participatory processes generate detailed case study responses for specific situations. Furthermore, they can be used to explore farm financial implications, investment strategies, barriers to adoption and potentially maladaptive practices. These processes, however, are resource-intensive and cannot be implemented in the numbers needed to include all the decision makers in the sector nor to allow robust scaling up of the results to national level for broader policy decisions. As discussed below, they also tend to focus adaptation effort on incremental technical responses within existing activities, potentially overlooking more transformative opportunities to adapt.

An alternative approach for national scale assessments is to use very general adaptation responses (see, for example, Heyhoe *et al.* 2007). These, however, lack the detail needed for local implementation and can be difficult to relate to realistic changes in farm management practices (Figure 9.3). Consequently, one middle way between these two extremes is to work with communities, industries and government to identify a diverse set of adaptation options and the general contexts in which each is appropriate (Figure 9.3). However, approaches of this kind need to be evolved via participatory engagement with communities, industries and governments in order to move beyond academic interest exercises towards empowering adaptation by decision makers.

One way of creating and implementing this approach is to combine local knowledge with systems modelling to provide a menu of options for different amounts of rainfall and temperature change (Figure 9.4). Adaptations proposed through implementation of this approach by farmers in the Brim region of Victoria have included reducing the proportion of higher risk crops grown (e.g. legumes); fallow on a regular basis on heavier soils (crop less); selecting for shorter season varieties and more heat-tolerant crops; reducing stocking rates and retaining more stubble to conserve soil moisture; considering growing more fodder crops, and reducing reliance on wheat cropping. Farmers can explore this menu at varying levels of detail, taking into account their knowledge, aspirations, attitudes, skills and practices.

The value of these types of within-system management adaptations to climate change could be substantial. For example, for the Australian wheat industry just two of these management adaptations (change in varieties and time of sowing to match the prevailing climate) could provide average annual benefits of between $150 million and $500 million per year by 2070 (Howden and Crimp 2005). The greatest benefits in this study occurred under conditions

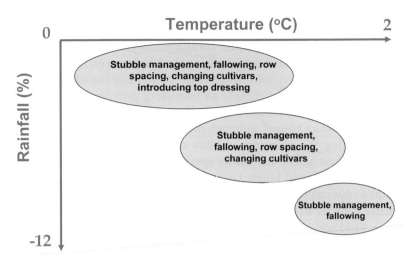

Figure 9.4: A simple typology of adaptation responses for cropping systems for varying scenarios of rainfall and temperature from Brim, Victoria.

of high climate change. Additional benefit is likely if a more comprehensive set of incremental adaptation options were to be considered (Howden *et al.* 2007). In the case study below, we extend this previous work to explore a significant, positive consequence of effective adaptation – the extent to which it will contribute to wheat exports, offsetting the anticipated effects of climate change and increased domestic demand arising from population growth and consumption patterns.

Climate change adaptation interacting with population growth and consumption patterns: a case study

Background

Wheat is the major crop in Australia. Production exceeds domestic consumption on such a scale that about 74% (15 million tonnes per annum (p.a.)) of the wheat produced each year is exported, earning about 4% of Australia's total exports, averaging $3.1 billion p.a. since 1990 (ABARE 2009). These exports contribute about 12% to the international trade in wheat. The extent to which Australia's wheat exports contribute to global food security depends on a complex set of social, economic and institutional factors beyond the scope of this chapter. We note, however, that to the extent that Australian exports firstly increase access to food by vulnerable groups and secondly, reduce the proportion of household budgets spent on food, other things being equal, higher exports can be said to contribute to greater global food security.

Wheat production is highly sensitive to climatic influences, with the difference between wheat production in a wet year and a dry year being almost a factor of three. Previous studies have indicated that the projected higher temperatures and lower rainfall across wheat-growing regions is likely to reduce yields. This is despite the generally positive effects on yields of elevated atmospheric carbon dioxide concentrations which increase water- and light-use efficiency (see, for example, Howden 2002). In this study, we assess the likely impact of projected changes in rainfall, temperature and elevated carbon dioxide on national wheat production and exports with and without farm-level adaptation.

production, for example, is expected to decline relative to other countries, potentially reducing the demand for Australian exports. This study, however, has highlighted the importance of placing climate change in the context of other key drivers as well as the critical importance and substantial benefit of developing effective adaptation options.

Limits to adaptation: incremental and transformational change

The example above showed that there could be significant benefits from even a small number of relatively simple, incremental changes to existing farming systems. Importantly, there are hundreds of such adaptations with many of these now starting to be identified and tested (Stokes and Howden 2009). However, the benefits of these types of adaptations are likely to plateau as more extreme climate changes are experienced over the coming decades. For example, practices such as zero till, stubble retention and crop canopy management will allow continued production even with substantial rainfall reductions. However, even these techniques will provide limited additional benefit in the face of more severe declines in rainfall. When expressed as the ratio of yields with adaptations compared to yields without adaptation (the benefit from adaptation), this typically shows a rapid increase in benefit with small amounts of climate change, but with this benefit levelling off with larger amounts of climate change (Figure 9.6). That is to say, there are limits to the effectiveness of incremental adaptations. This response has been found in both national responses for wheat (Howden and Crimp 2005) and with a global meta-analysis of crop responses (Howden *et al.* 2007). Interestingly, in both cases, the plateau started at around 2°C temperature rise, although the point at which the adaptation benefit levels off is rainfall-dependent (Figure 9.6).

The existence of limits to the benefits of incremental adaptations raises the question as to how Australian agriculture may adapt under the scenarios of much greater change than the 2°C indicated in the analyses above. A 0.9°C temperature rise has already been experienced,

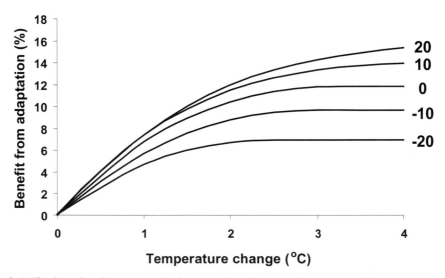

Figure 9.6: The benefit of incremental adaptation for the Australian wheat industry (expressed as % increased in yield from the case with no management adaptation) with increased temperature and various scenarios of rainfall change. (Modified from Howden and Crimp 2005)

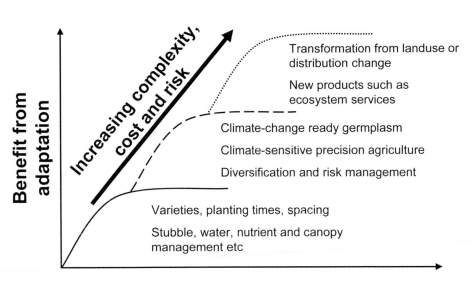

Figure 9.7: Hypothesised relationship between incremental and more transformational adaptations as climate change increases, indicating possible types of adaptations and the likely increasing complexity, cost and risk associated with the more transformative adaptations.

meaning that Australia is already almost halfway to the 2°C change. We suggest that there are a range of more transformative changes that could be considered in addition to incremental adaptations of existing agricultural systems. For example, there are options for more substantial adaptations such as development of more climate change-adapted crops and livestock, more effective integration of climate into precision-agriculture operations and broad-scale diversification of farm household incomes (Figure 9.7). In the event that the climate changes even further, there may be more transformative options such as major alteration of land use at a location or change in the location of significant industries or diversification of products such as provision of ecosystem services including carbon sequestration.

These changes are not necessarily sequential. For example, there are already instances in Australia where industries are reported to be making transformative changes primarily as a result of climate changes. These examples include 1) the development by the Peanut Company of Australia of areas in the Northern Territory away from their historical base in south-eastern Queensland; 2) the adoption of cropping in the high rainfall grazing zones of southern Australia which were previously too wet for cropping, and 3) the search for alternative rice-growing locations following extended periods of very low water allocations in the irrigation areas of south-eastern Australia. As we move from incremental to transformational changes, however, it is likely that the complexity, cost and risk of the changes increases (Figure 9.7). There are likely to be several key roles for science in terms of supporting transformational change. These include helping design robust agricultural systems which meet triple bottom-line goals, provision of the underlying technological options for farmers to use (e.g. appropriate genetic material) and risk assessments associated with their introduction, policy research into institutional settings that may impede or support transformation and social research to understand the transformation processes in both the groups that transform and those that choose not to. This research may enable us to avoid repeating some of the mistakes of past large-scale agricultural developments.

10. WHEAT, WINE AND PIE CHARTS: ADVANTAGES AND LIMITS TO USING CURRENT VARIABILITY TO THINK ABOUT FUTURE CHANGE IN SOUTH AUSTRALIA'S CLIMATE

Peter Hayman and Bronya Alexander

Abstract

One of the costs of climate change is that it starts to erode the value of climate knowledge developed by farmers and farm advisers over decades. Paradoxically, this local contextualised knowledge of managing climate variability is one of the most valuable assets in managing the coming decades of climate change. Rather than encouraging farmers to ignore their experiential knowledge of climate, the challenge is to find appropriate ways to use their understanding of climate variability to manage climate change better.

The Australian Bureau of Meteorology uses deciles to quantify and communicate climate variability. Farmers and their advisers commonly use deciles to express the risk associated with climate variability. A decile 3 for growing season rainfall communicates that in the historical record, seven years have been wetter and two drier. Although not as common, the same approach can be used for growing season temperature in viticulture. Climate change science tends to refer to per cent change in mean rainfall and change in average temperature. In this chapter we show how we have used the revised probability of deciles with dryland farmers and wine grape growers in South Australia to examine the effects of and adaptation to climate change in coming decades.

Introduction

Australian farmers have shown a high level of adaptive capacity in a variable climate. Dealing with the early stages of climate change is largely an extension of this.

One of the consequences of climate change, however, is that the experiential knowledge of local climates starts to become less valuable (Quiggen and Horrowitz 2001). Although difficult to assess, there is considerable value in the understanding accrued over decades and generations of dryland farmers contained in part from rainfall records but more importantly, in knowing the risks and opportunities presented by dry years, the wet years and those in between. Likewise in the wine grape industry, there is invaluable local knowledge in the vineyard and winery about dealing with warmer vintages and cooler vintages. This presents a paradox; on the one hand climate change diminishes the value of local knowledge while on the other, this local knowledge is the most valuable asset for managing the coming decades of a variable and changing climate.

In this chapter we address an aspect of this paradox by describing one way of using knowledge about current variability to manage future change. The chapter is organised as follows: we start by clarifying terms such as 'climate variability' and 'climate change' and provide a brief background to climate risk in dryland farming and wine grape production in South Australia. We then review some of the challenges of communicating uncertain climate information, and then describe a method that we have found useful in talking with dryland farmers and wine grape growers in South Australia.

For the purposes of this chapter the essential difference between managing for climate variability and managing for climate change relates to assumptions about whether climate can be assumed to be stationary over the planning horizon. There has long been a discussion of managing climate variability in Australia based on the observation that Australian agriculture deals with a high degree of year-to-year variability due to internal drivers within the climate system such as the El Niño–Southern Oscillation, the Indian Ocean Dipole, the Southern Annular Mode and the Interdecadal Pacific Oscillation (CSIRO and Bureau of Meteorology 2007). Independent of climate science identifying the source of the internal variability, agriculturalists have long recognised that climate is variable, even cyclical, but over a long enough period, assumed to be stationary. In other words, there is the assumption that fluctuations are bound within an envelope of defined variability.

Climate change involves planning for a non-stationary climate. Milly *et al.* (2008) observed that the concept of accepting that climate is not stationary challenged the basis of planning and risk assessment that permeated training and practice for water engineers. The same is true for most agricultural scientists, policy makers and farmers. An interesting policy case is that of drought and exceptional circumstances where different base periods can lead to quite different assumptions about how 'exceptional' a given event may be. Projected changes in temperature and rainfall are likely to make what was seen as exceptional become more common and in the case of temperature, normal or cool (Hennessy *et al.* 2008).

This chapter focuses on the near future of the coming two to three decades and the projections that are widely available from CSIRO and Bureau of Meteorology (2007). There is little doubt that there are levels of climate change projected for later in the century (that may occur earlier) that will be very difficult for Australian farmers to adapt to without transformational change (Stokes and Howden 2008).

Climate risk in dryland farming and wine grape growing

Dryland farming (annual gross production A\$2.3 billion (B)) and wine grape production (annual gross production A\$1 B) (PIRSA 2009) are important to the South Australian (SA) economy and vital to rural communities. Both cereal grain production and irrigated wine grape growing industries are adapted to a Mediterranean climate but have different sensitivities; dryland farming is most sensitive to year-to-year changes in precipitation, and wine grape production is more sensitive to temperature. There is also a range in adaptive capacity and importantly, in planning horizons between annual crops and perennial crops (Webb and Barlow 2008). Like most agricultural industries, this sector of the SA economy is exposed and sensitive to climate. South Australia is the driest state in the driest inhabited continent and draws irrigation water at the downstream end of the Murray–Darling Basin system which has had the lowest inflows on record (MDB 2009).

Recent seasons have highlighted the vulnerability of dryland and irrigated farming in South Australia. The 13-year period from October 1996 to September 2008 was the driest on record for the southern areas of the cropping belt and well below average rainfall for the rest of the cropping belt (BoM 2008). Wine grape growers, like other irrigators, have experienced substantial cuts to their water allocation. They have also experienced damaging heat waves in February 2004, March 2008 and January 2009. Analysis of recent crop development data shows that grapevines are maturing earlier in the season (Petrie and Sadras 2008). One of the implications of the earlier maturity is that the sensitive ripening period is shifted from cooler autumn conditions to the hotter conditions of late summer (Webb and Barlow 2009).

Despite adverse climate events, it is important to acknowledge that there is a high level of adaptive capacity in the grains and wine grape industry. Black *et al.* (2008) estimated a wheat yield efficiency gain of 1.7% p.a. from 1977 to 2006 with a slight but significant reduction in the rate of gain towards the end of the period. Attribution of the plateau in efficiency gains is inconclusive. In viticulture, Sadras *et al.* 2007 used vintage scores over the last three decades to show that despite warming trends, water shortages and financial pressure, there is an increase in both vintage quality and a reduction in variability in quality from vintage to vintage. There are many in the wine industry able to provide clear examples of innovations which minimised the damage from recent heatwaves (Hayman *et al.* 2009).

Dryland farmers and viticulturists in South Australia have an understanding of the year-to-year variability and consider this in their decision making. Successful managers tend to find options that minimise the disadvantages and maximise the benefits of variability. As pointed out by farmers in low rainfall environments, variability is a source of opportunity as well as downside risk (Mudge 2008). The longer-term climate of a region determines whether wine grapes can be successfully grown or not. If grapes can be grown, it dictates the appropriate varieties and the style of wine for that region. The quality characteristics of an individual vintage are largely an expression of the climate and weather events of that season.

Understanding variability with deciles

The Australian Bureau of Meteorology provides an extensive range of information on weather and climate to agricultural industries. One of the ways that the variability in seasons is analysed and communicated is with the use of deciles whereby the recent season (three months or six-month period) is ranked against all previous seasons. This is effectively communicated in maps on the Bureau of Meteorology website and is a common topic of discussion amongst farmers and advisers. The use of deciles is in part due to the fact that rainfall is rarely normally distributed (Gibbs and Maher 1957).

The Bureau of Meteorology also uses deciles for temperature and this is communicated through maps. It is straightforward to communicate that a cool vintage (say 2001) had a growing season temperature of decile 1. As we will show, 2001 was an exceptionally cool year in a warming trend. Most of the vintages since 2001 have been decile 10. There are many indices that can be used to link climate to viticulture, including Growing Season Temperature (GST) – October to April in southern hemisphere – which is commonly used internationally. In their analysis of 27 quality wine regions in the world, Jones *et al.* (2005) listed three Australian regions, the Margaret River (GST 18.6°C), the Hunter Valley (GST 19.8°C) and the Barossa (GST 19.9°C), as the 21st, 23rd and 24th warmest regions respectively. The bulk of Australian wine is produced in the hot inland regions (GST >20°C) using irrigation water from the Murray or Murrumbidgee rivers.

Cereal growing in South Australia is largely dependent on winter rainfall often expressed as growing season rainfall (GSR) which is the sum of rainfall between April and October. A robust relationship between GSR and yield established by French and Schulze (1984) is commonly used by farmers and advisers, where 110 millimetres (mm) subtracted from GSR and the resulting value in mm is multiplied by a water use efficiency term of 20 kg/mm of rainfall to estimate the yield (kg of grain per hectare). Hence expressing GSR as deciles can be related to yield. A decile 3 season is understood as meaning two seasons out of 10 would be expected to be drier (and lower yielding) and seven wetter.

The challenge of communicating worrying but uncertain information from climate science

Table 10.1 shows the projections for South Australia. As with most climate change projections, these show a relatively narrow range for temperature and a wide range for rainfall. For many agricultural enterprises in South Australia, any shift towards a warmer and drier future is worrying but the wide range of projections indicate the uncertainty. Not only are the projections uncertain, it is difficult to conceive what a degree rise in temperature means, or how a 10% decline in rainfall might change the risk profile. The question of what a 10% decline in rainfall means in terms of shifts in deciles is not simple to answer. Given that farmers and agronomists use deciles as their expression of risk, there is a communication mismatch.

Table 10.1 shows a greater degree of inter-model agreement for temperature than rainfall. The range for rainfall can be characterised as uncertain. Given the uncertainty at a local level, especially for rainfall, Stokes and Howden (2008) recommend a risk-based approach that focuses on a range of plausible impacts rather than potentially misleading average, median or even most likely projection. In their submission to the Parliamentary Inquiry into farmers and climate change, the Australian Institute of Agricultural Science cautioned against an over-emphasis on modelling the exact nature of future climate change and recommended a range of plausible scenarios that could be used to assess the capacity of industries to adjust (AIAST 2009).

Patt and Dessai (2005) reviewed the many challenges of communicating uncertain projections of climate change. These difficulties are similar to those experienced in communicating seasonal climate forecasts (Nicholls 1999), and the large amount of psychology literature from fields as diverse as medicine, engineering, investment and legal judgements which shows that most people struggle with interpreting probabilities (Burnstein 1996). Patt and Dessai (2005) argued that natural stochastic uncertainty due to the chaotic nature of climate system, and epistemic uncertainty due to incomplete knowledge of climate processes, can be quantified by traditional uncertainty analysis such as Monte Carlo and ensemble runs. This is represented in the 10th, 50th and 90th percentile of model runs in Table 10.1. They distinguished this sort of uncertainty with uncertainty due to human behaviour and technology advances on emissions which can't be meaningfully quantified (the different emission scenarios in Table 10.1). They recommended a scenario analysis (Patt and Dessai 2005).

Marx *et al.* (2007) maintained that a major challenge for communicating climate information was that those preparing the information tended to assume that decision makers processed the information analytically, whereas many studies showed that people were more likely to rely heavily on experiential processing systems. They recommended a linking of statistical information to concrete experiences as a way of making sure that the users understood the information and could use it for planning. As an historian, Sherratt (2005) observed that we cannot reliably remember climate because memory generates meaning rather than statistics.

A simple approach with time series and pie charts

In the following section we describe a simple approach to communicating what a projected change in climate would mean in terms of current climate variability. This information is based

Table 10.1 Climate change projections for Adelaide, for 2030 using the SRES medium emissions scenario (A1B) and for 2070 using a low (B1) and high (A1FI) emissions scenario. Projections are relative to the base period of 1980–1999 and show the spread between the 10th, 50th and 90th percentile (p) of 23 climate models (see CSIRO and Bureau of Meteorology 2007, p. 50)

	Season	2030 A1B 10p	2030 A1B 50p	2030 A1B 90p	2070 B1 10p	2070 B1 50p	2070 B1 90p	2070 A1FI 10p	2070 A1FI 50p	2070 A1FI 90p
Temperature (°C)	Annual	0.6	0.9	1.3	1	1.5	2.1	1.9	2.8	4
Rainfall (%)	Annual	–11	–4	2	–18	–7	4	–32	–13	8
	Summer	–14	–2	11	–23	–3	18	–39	–5	35
	Autumn	–11	–1	9	–18	–2	14	–31	–4	28
	Winter	–15	–6	2	–23	–10	3	–40	–19	6
	Spring	–19	–8	3	–30	–12	4	–50	–23	8

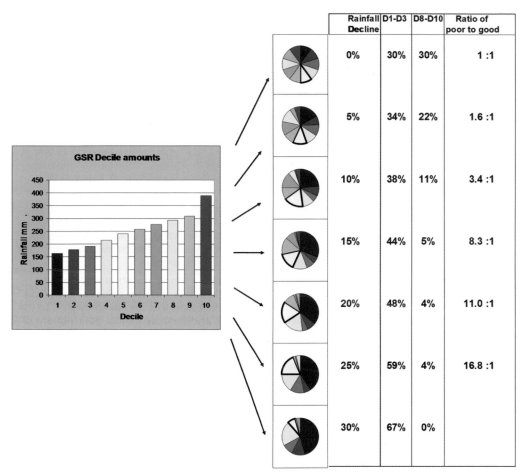

Rainfall Decline	D1-D3	D8-D10	Ratio of poor to good
0%	30%	30%	1 :1
5%	34%	22%	1.6 :1
10%	38%	11%	3.4 :1
15%	44%	5%	8.3 :1
20%	48%	4%	11.0 :1
25%	59%	4%	16.8 :1
30%	67%	0%	

Figure 10.3: Pie chart or probability wheels showing outcomes for deciles from Figure 10.1 (current climatology) in a range of drying scenarios. Colour scheme for deciles is the same as Figure 10.1.

reduction in rainfall. For example, in low rainfall farming areas, a reduction in rainfall of 15% results in a 50% chance of being in the lowest three deciles and only a 10% chance of being in the top three deciles. Furthermore the chance of what are now considered two poor seasons in

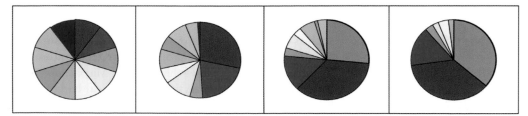

Figure 10.4: Pie chart or probability wheels showing outcomes for deciles from Figure 10.2 (1960 to 1990) climatology in a range of warming scenarios. Colour scheme for deciles is the same as Figure 10.2 with conditions warmer than ever experienced in the climatology shown in grey. From left to right, pie chart representing no change, 0.5, 1.0 and 1.5°C rise in temperature.

a row (decile 3 or below) increases from 9% in the current climate to 25% in the future climate or from one chance in 10 to one chance in four.

Concluding remarks

IPCC Working Group 2 (2007) suggested that adaptation assessments would benefit from linking future changes in climate to past and present climates. We have presented an approach to link the near future with the present and recent past climate. This approach has limits, perhaps the most obvious being that any focus on adaptation in the near future can ignore the mid-term future (say, after 2050) when past knowledge and expectations of a local climate may be a liability rather than an asset. Stokes and Howden (2008) note that managing shorter term variability over coming decades is likely to be most effective when the trend in the shorter term variability is the same as the longer term change. If the trend and the projection are in opposite directions and the projection turns out to be correct, there is a source of maladaptation. They cite the case of central Western Australia where the projections are for drying yet the trend over past decades has been wetting. The method that we have described could be used to highlight this difference.

The method as we have described is appropriate for growing season rainfall or growing season temperature rather than extreme weather events such as heatwaves, frosts or excessive rain. Not only are extreme events more difficult to categorise, the damage is very sensitive to the timing of the event at key crop stages. Another limit to the approach is that although we have used a grey section of the probability wheel to indicate the chance of being warmer than any previous season, the method also does not take into account the likelihood of novel climates which have no analogue in time or space (Williams *et al.* 2007).

When we have conducted these exercises with dryland farmers and wine grape growers, we have found that they are able to articulate the challenges and opportunities from a changing climate. These discussions give good reason to be optimistic about the ability of agriculture to adapt to early and milder projections. The same methodology, with the same participants, gives reason for alarm at higher levels of climate change.

References

AIAST (2009) 'Australian Institute of Agricultural Science and Technology (AIAST) Submission to Parliamentary Inquiry into the role of governments in assisting farmers adapt to the impact of climate change. Submission Number 63.' AIAST, Canberra. <www.aiast.com.au/index.php?menu=communications&action=pub_submission>

Black I, Dyson C, Hayman P and Alexander B (2008) The determinants of South Australian wheat yield increases. 14th Australian Society of Agronomy conference, 21–25 September 2008, Adelaide, SA. <www.regional.org.au/au/asa/2008/concurrent/evaluating_systems/5837_blackid.htm>

BoM (Bureau of Meteorology) (2008) Special Climate Statement 16. Australian Bureau of Meteorology. <www.bom.gov.au/climate/current/statements/scs16.pdf>

Bernstein PL (1996) *Against the Gods, The Remarkable Story of Risk*. John Wiley, New York.

CSIRO and Bureau of Meteorology (2007) 'Climate change in Australia – technical report 2007'. CSIRO and Bureau of Meteorology, Australia. <www.climatechangeinaustralia.gov.au>

French RJ and Schulze JE (1984) Water use efficiency of wheat in a Mediterranean environment. I The relation between yield, water-use and climate. *Australian Journal of Agricultural Research* **35**, 734–764.

Gibbs WJ and Maher JV (1957) Droughts in Australia: Bulletin No. 43. Melbourne, Commonwealth Bureau of Meteorology.

Hayman PT, McCarthy M and Grace WG (2009) Assessing and managing the risk of heatwaves in South-Eastern Australian winegrowing regions. *The Australian and New Zealand Grapegrower and Winemaker* **543**, 22–24.

Hennessy K, Fawcett R, Kirono D, Mpelasoka F, Jones D, Bathols J, Whetton P, Stafford Smith M, Howden M, Mitchell C and Plummer N (2008) 'An assessment of the impact of climate change on the nature and frequency of exceptional climatic events'. CSIRO and Australian Bureau of Meteorology, Melbourne.

IPCC (2007) Climate Change 2007: Impacts, Adaptation and Vulnerability. Contribution of Working Group II to the Fourth Assessment Report of the Intergovernmental Panel on Climate Change. Cambridge University Press, Cambridge, UK.

Jones GV (2006) Climate change and wine: observations, impacts and future implications. *Wine Industry Journal* **21**, 21–26.

Jones GV, White MA, Cooper O and Storchmann K (2005) Climate change and global wine quality. *Climatic Change* **73**, 319–343.

Marx SM, Weber EU, Orlove BS, Leiserowitz A, Krantz DH, Roncoli C and Phillips J (2007) Communication and mental processes: experiential and analytic processing of uncertain climate information. *Global Environmental Change-Human and Policy Dimensions* **17**, 47–58.

Murray Darling Basin Authority (2009) Drought Update. Issue 19 June 2009. <www.mdba.gov.au>

Milly PCD, Betancourt J, Falkenmark M, Hirsch RM, Kundzewicz ZW, Lettenmaier DP and Stouffer RJ (2008) Climate change: stationarity is dead: whither water management? *Science* **319**, 573–574.

Mudge B (2008) Managing risk in an uncertain climate. 14th Australian Society of Agronomy conference, 21–25 September 2008, Adelaide, SA. <www.regional.org.au/au/asa/2008/plenary/farming_uncertain_climate/6313_mudgeb.htm>

Nicholls, N (1999) Cognitive illusions, heuristics and climate prediction. *Bulletin of the American Meteorological Society* **80**, 1385–1397.

Patt A and Dessai S (2005) Communicating uncertainty: lessons learned and suggestions for climate change assessment. *Comptes Rendus Geoscience* **337**, 425–441.

Petrie PR and Sadras VO (2008) Advancement of grapevine maturity in Australia between 1993 and 2006: putative causes, magnitude of trends and viticultural consequences. *Australian Journal of Grape and Wine Research* **14**, 33–45.

PIRSA (2009) Primary Industries and Resources South Australia. South Australian Food Industries Overviews. PIRSA, Adelaide. <http://outernode.pir.sa.gov.au/foodSA/scorecard/industry_scorecards>

Quiggin J and Horowitz J (2001) 'Costs of Adjustment to Climate Change'. Working Papers in Economics and Econometrics. Working Paper No. 415. The Australian National University.

Sadras VO, Soar CJ and Petrie PR (2007) Quantification of time trends in vintage scores and their variability for major wine regions of Australia. *Australian Journal of Grape and Wine Research* **13**, 117–123.

Sherratt T (2005) Human elements. In: *A Change in the Weather: Climate and Culture in Australia*. (Eds T Sherrat, T Griffiths and L Robin) pp. 1–17. National Museum of Australia Press, Canberra.

Stokes CJ and Howden SM (Eds) (2008) An overview of climate change adaptation in the Australian agricultural sector. CSIRO. <www.csiro.au/resources/AgricultureAdaptationReport2008.html>

Webb LB and Barlow EWR (2008) Viticulture. In *An overview of climate change adaptation in the Australian agricultural sector*. (Eds CJ Stokes, SM Howden) pp. 158–171. CSIRO Publishing, Melbourne. <www.csiro.au/resources/AgricultureAdaptationReport2008.html>

Williams JW, Jackson ST and Kutzbach JE (2007) Projected distributions of novel and disappearing climates by 2100 AD. *Proceedings of the National Academy of Science* **104**, 5738–5742.

11. MANAGING EXTREME HEAT IN THE VINEYARD: SOME LESSONS FROM THE 2009 SUMMER HEATWAVE

Leanne Webb, John Whiting, Andrea Watt, Tom Hill, Fiona Wigg, Greg Dunn, Sonja Needs and Snow Barlow

Abstract

In January and February 2009, a severe heatwave affected much of the south-eastern region of Australia causing heat stress-related damage to many vineyards. A survey of the impact of the heatwave in 10 wine grape-growing regions was undertaken to determine the extent of the damage and the management strategies used to ameliorate the effects of the heatwave. The most valuable information revealed as a result of this survey was that the variation of impact was *not* related to the level of temperature to which the vines were exposed. Variation in the management strategies, either traditional approaches or reactive management, has highlighted potential best practice methods for dealing with extreme heat events in future. Generally there was more reported damage where water was not available, rows were oriented in a north–south direction, berry development was between the veraison and harvest stages, and bunches were exposed to radiation. Implementation of identified management options to ameliorate the impact from extreme events can be introduced at different phases of vineyard development: at planning stage, inter-seasonal or within a season.

Introduction

An exceptional heatwave occurred in south-eastern Australia during late January and early February 2009. Many records were set for high day- and night-time temperatures as well as for the duration of extreme heat (National Climate Centre 2009). At this time most of the south-eastern Australian wine grape crop was in the veraison (berry softening and commencement of sugar accumulation), or post-veraison stage of its phenology (McIntyre *et al.* 1982). The effects of the heatwave on vineyards appear to be unprecedented with significant heat stress-related crop losses at some sites.

Global average surface temperature has increased by approximately 0.7°C since the beginning of the 20th century. The warming has been associated with more heatwaves, changes in precipitation patterns, reductions in sea ice extent and rising sea levels globally (IPCC 2007). Increases in heatwave occurrences in eastern Australia and South Australia since the 1950s have been measured (Deo *et al.* 2007) and are projected to increase in the future (Alexander and Arblaster 2009).

By conducting a study of intra- and inter-regional variation of wine grape vineyard impact and management response associated with the 2009 heatwave in south-eastern Australia, successful management options were revealed that significantly reduced the impact. This survey of real-time responses to a severe heatwave supports and validates recommendations for management options identified during some recent workshops held in Australia (Hayman *et al.* 2009), and at the same time quantifies the success of such strategies (Webb *et al.* 2009). As increases in the frequency of extreme hot days (>35°C) in the growing season are projected to eliminate wine grape production in many areas of the United States (White *et al.* 2006), these practices may become more critical to ongoing profitable wine production in this country and globally.

The heatwave

According to the National Climate Centre (2009) there were two major periods of exceptional high temperatures; 28–31 January (Figure 11.1) and 6–8 February (Figure 11.2). On 27–31 January in southern South Australia, and much of central, southern and western Victoria, maximum temperatures throughout reached their highest levels since at least 1939. The extreme heat on the 7 February, where record high temperatures for February were set over 87% of Victoria, also affected the southern fringe of New South Wales and eastern South Australia. Renmark in the Riverland wine grape-growing region set a February record (48.2°C). In addition to its peak intensity, the heatwave was also notable for its duration with slightly lower, but still very high, temperatures persisting in many inland areas through the intervening period (1–5 February).

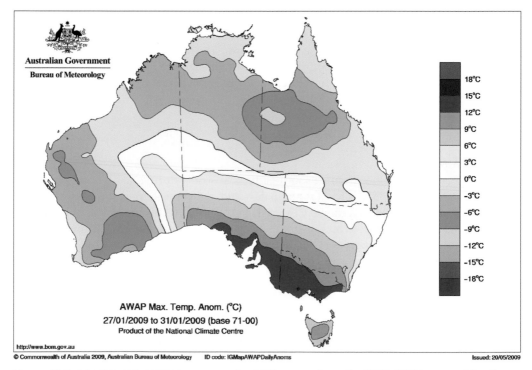

Figure 11.1: Maximum temperature anomalies (differences from the 1971–2000 average) for the period 27–31 January 2009. (Map supplied by Australian Bureau of Meteorology)

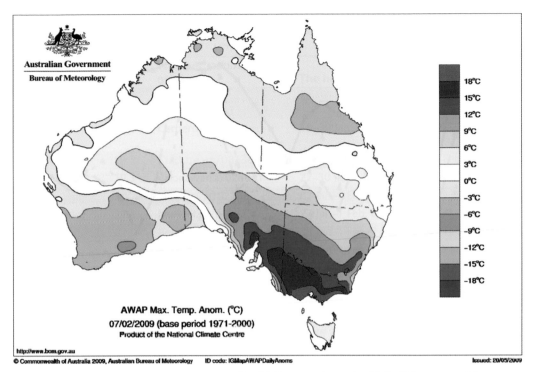

Australian Government
Bureau of Meteorology

18°C
15°C
12°C
9°C
6°C
3°C
0°C
-3°C
-6°C
-9°C
-12°C
-15°C
-18°C

AWAP Max. Temp. Anom. (°C)
07/02/2009 (base period 1971-2000)
Product of the National Climate Centre

http://www.bom.gov.au
© Commonwealth of Australia 2009, Australian Bureau of Meteorology   ID code: IGMapAWAPDailyAnoms   Issued: 20/05/2009

Figure 11.2: Maximum temperature anomalies (differences from the 1971–2000 average) for 7 February 2009. (Map supplied by Australian Bureau of Meteorology)

Across the winegrowing regions, the heatwave varied in intensity, duration and also diurnally as can be seen by comparing records from Mildura (Murray–Darling/Swan Hill region) where 12 consecutive days above 40°C occurred, at Launceston (Tasmania) where one spike in temperature was observed, and Cerberus (Mornington Peninsula) with two heat-spikes (Figure 11.3).

The heatwave was also accompanied by very dry conditions, both during and in the weeks leading up to the event. Further, the south-eastern region of Australia has been subject to a protracted drought with below-average rainfall experienced for the past 12 years (Australian Bureau of Meteorology 2008). Not only were soil moisture reserves low in winegrowing regions without access to public irrigation schemes, but growers may have been influenced not to irrigate because of reduced allocations of irrigation water from these schemes or because of reduced on-farm water storages (Webb *et al.* 2008).

On the 7 February, catastrophic bushfires affected some of the wine grape-growing regions that were also affected by the heatwave (Karoly 2009). While smoke from bushfires can adversely affect wine grapes (Kennison *et al.* 2007), this study focused solely on heat impacts to vineyards.

The survey

Assessment of vineyard responses and impact was undertaken by interviewing managers of properties from selected winegrowing regions located within the affected areas: Mornington

Figure 11.5: Examples of some of the impacts of the 2009 summer heatwave observed across the winegrowing regions of south-eastern Australia. (a) Shoot tip burn (Pinot Noir vines), (b) leaf burn (Shiraz), (c) sunburn on Chardonnay berries, (d) stalled development (Pinot Noir), (e) 'bagging' and desiccation of berries, (f) Shiraz berry desiccation.

from each region (typically <5–10% of the total), so while attempts were made to gain an unbiased overview of the impact by keeping the sample random, the outcomes should be considered within this context.

Results

Nearly everyone reported they had ample warning of the event and were satisfied with the weather information available to them. In a few cases the managers found the event was either more severe or of longer duration than expected, even given the forecast. Where two distinct hot temperature spikes were recorded, it was often during the initial spike that most of the damage occurred. Most believed that the combination of the intensity of the heat *and* the duration contributed to the impact.

The extent of the damage (as estimated by the vineyard managers) was found to vary between regions, within regions and within vineyards. In this preliminary assessment the estimates of average percentage and severity of damage (in a region) were considered and subjective judgement was used to categorise the regional damage estimates into worst affected, medium impact, least affected and negligible impact (Figure 11.6). Subsequent quantitative analysis has been be undertaken (Webb *et al.* 2009) to confirm this preliminary assessment.

Where relatively higher levels of damage were reported in the properties surveyed (e.g. Mornington Peninsula, McLaren Vale and Rutherglen), several circumstances were reported.

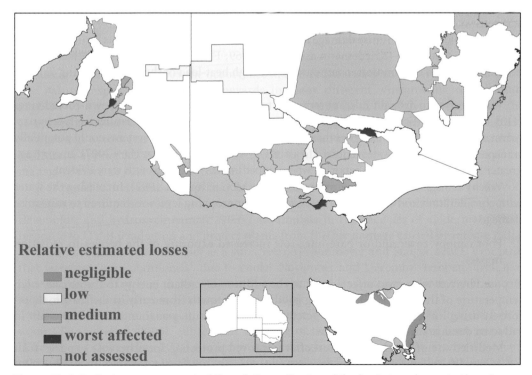

Figure 11.6: Preliminary assessment of the relative estimates of the heat stress losses to the wine grape crop in south-eastern Australia.

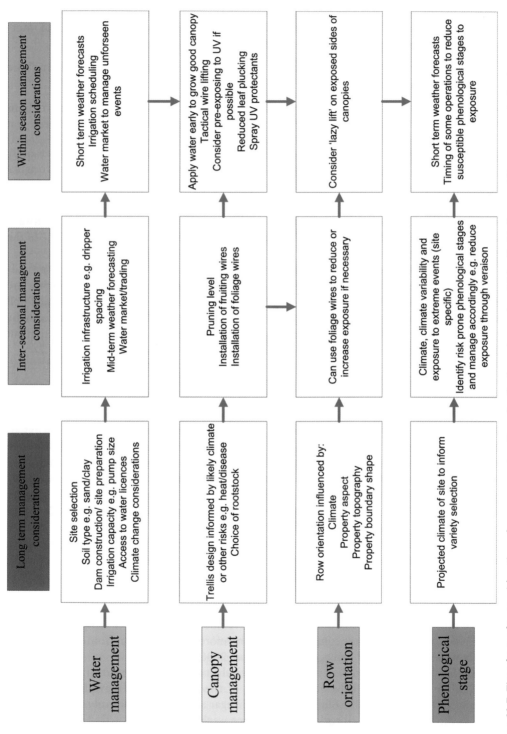

Figure 11.7: Time frame for consideration of management options for the four vineyard variables that influenced the response to heatwaves.

acclimatisation may have contributed to these increased delays in ripening, as consistently higher temperatures throughout a growing season has been shown to produce more stable photosynthetic capacity in grapevine leaves (Zsofi *et al.* 2009).

Significant night-time stomatal conductance and transpiration is associated with higher daytime evapotranspiration values and the driving gradient between the leaf and the atmosphere at night (Snyder *et al.* 2003). Warm nights were experienced in all regions implying possible night-time water loss. McLaren Vale, a region with reported high levels of impact, experienced exceptionally high overnight temperatures which may have increased water demand and, if not factored in to the irrigation budget, contributed to the heat-stress. While no studies have yet reported or quantified water loss at night in grapevines, with warm nights projected to increase (Alexander and Arblaster 2009) it may be prudent to re-evaluate water budgeting to incorporate night-time losses.

Other useful strategies mentioned by the managers are noted for consideration and perhaps further investigation:

- grapevine nutrition and overall health were deemed important to withstanding the added stress of an extreme heat event;
- physical barriers to UV radiation, e.g. kaolin clay sprays used in apple production (see Glenn *et al.* 2001) require further evaluation for the wine grape industry, and
- having diesel back-up for electric irrigation pumps can be important due to the increased risk of power cuts during severe heat events.

Across the 10 regions surveyed there is a range of management practices employed with regard to fruit exposure, as well as varied levels of sophistication of water management. These practices have evolved in response to the risks to which the vineyards are normally exposed, such as high temperatures in the northern vineyards and disease pressure in the more southern vineyards. It must be emphasised that a major shift in management in response to this heatwave (e.g. reduction to fruit exposure) may predispose the crop to other risks in more typical seasons, with similar magnitudes of losses. Adjusting viticultural management will be, in the future even more than previously, a risk minimisation exercise. Questions of fruit exposure versus non-exposure, the cost of water security, rootstock, vine training and trellising, and choice of row orientation may need to be re-evaluated being mindful of typical climate variability, but with due consideration of a hotter projected climate.

Conclusion

Capturing the observations and management decisions made by a cross-section of wine grape industry practitioners soon after a severe heatwave has proved extremely informative. Their depth and breadth of knowledge, both intuitive and technical, has revealed diverse approaches that have ameliorated the impact of the heatwave. Documenting these, as their effectiveness varied across and within regions and even vineyards, allows the industry to identify the management strategies that may assist this important sector and other horticultural enterprises to cope in an increasingly challenging environment.

Acknowledgements

We wish to thank all the vineyard managers who so willingly participated in the interviews during a busy time of the season. We also wish to thank Graeme Russell (Macquarie

University) and Adam Peitch (CCW, Riverland) who assisted with development of the survey questionnaire. We also wish to acknowledge Ian Smith (CSIRO Marine and Atmospheric Research) and Mark Stafford Smith (CSIRO Climate Adaptation Flagship, Gungahlin) for their thorough review of former drafts of this document. Thanks also to Grape and Wine Research and Development Corporation (GWRDC) for funding the project.

References

Alexander LV and Arblaster JM (2009) Assessing trends in observed and modelled climate extremes over Australia in relation to future projections. *International Journal of Climatology* **29,** 417–435. <http://dx.doi.org/10.1002/joc.1730>

Australian Bureau of Meteorology (2008) 'Special Climate Statement 16: long-term rainfall deficiencies continue in southern Australia while wet conditions dominate the north. 10th October 2008.' <www.bom.gov.au/climate/current/statements/scs16.pdf>

Deo RC, McAlpine CA, Syktus J, McGowan HA and Phinn S (2007) On Australian heat waves: time series analysis of extreme temperature events in Australia, 1950–2005. In: *MODSIM 2007 International Congress on Modelling and Simulation. Modelling and Simulation Society of Australia and New Zealand.* (Eds Oxley L and Kulasiri D) pp. 626–635.

Downey MO, Dokoozlian NK and Krstic MP (2006) Cultural practice and environmental impacts on the flavonoid composition of grapes and wine: a review of recent research. *Americal Journal of Enology and Viticulture* **57,** 257–268.

Dry N (2007) *Grapevine Rootstocks. Selection and Management for South Australian Vineyards.* Lythrum Press, in association with Phylloxera and Grape Industry Board of South Australia, Adelaide, SA.

ESRI (2007) 'ArcINFO GIS mapping software'. <www.esri.com/software/arcgis/arcinfo/about/features.html>

Flexas J, Bota J, Cifre J, Escalona JM, Galmes J, Gulias J, Lefi E-K, Martinez-Canellas SF, Moreno MT, Ribas-Carbo M, Riera D, Sampol B and Medrano H (2004) Understanding down-regulation of photosynthesis under water stress: future prospects and searching for physiological tools for irrigation management. *Annals of Applied Biology* **144,** 273–283.

Frohnmeyer H and Staiger D (2003) Ultraviolet-B radiation-mediated responses in plants. Balancing damage and protection. *Plant Physiology* **133,** 1420–1428. <www.plantphysiol.org>

Glenn DM, Puterka GJ, Drake SR, Unruh TR, Knight AL, Baherle P, Prado E and Baugher TA (2001) Particle film application influences apple leaf physiology, fruit yield, and fruit quality. *Journal of the American Society for Horticultural Science* **126,** 175–181. <www.ars.usda.gov/SP2UserFiles/person/2017/Particle%20film%20application%20influences%20apple%20leaf%20etc.pdf>

Hayman P, McCarthy MG and Grace W (2009) Assessing and managing the risk of heatwaves in SE Australian wine regions. *Australian and New Zealand Grapegrower and Winemaker* **543,** 22–24.

IPCC (2007) Climate Change 2007: The Physical Science Basis. Contribution of Working Group I to the Fourth Assessment Report of the Intergovernmental Panel on Climate Change. Cambridge University Press, Cambridge, UK.

Karoly DJ (2009) The recent bushfires and extreme heatwave in southeast Australia. *Bulletin of the Australian Meteorological and Oceanographic Society* **22,** 10–12.

Kennison KR, Wilkinson KL, Williams HG, Smith JH and Gibberd MR (2007) Smoke-derived taint in wine: effect of postharvest smoke exposure of grapes on the chemical composition and sensory characteristics of wine. *Journal of Agriculture and Food Chemistry* **55,** 10897–10901.

Kliewer WM and Dokoozlian NK (2001) Leaf area/crop weight ratios of grapevines: influence on fruit composition and wine quality. In: *Proceedings of the ASEV 50th Anniversary Annual Meeting.* Seattle, Washington, June 19–23. (Eds Rantz JM) pp. 285–295. American Society of Enology and Viticulture.

Kriedemann PE and Smart RE (1969) Effects of irradiance, temperature and leaf water potential on photosynthesis of vine leaves. *Photosynthetica* **5,** 15–19.

McCarthy MG, Jones LD and Due G (1992) Irrigation principles and practices. In: *Viticulture Volume 2: Practices.* (Eds Coombe BG and Dry PR) pp. 104–128. Winetitles, Adelaide.

McIntyre GN, Lider LA and Ferrari NL (1982) The chronological classification of grapevine phenology. *American Journal for Enology and Viticulture* **33,** 80–85.

National Climate Centre (2009) Special climate statement: the exceptional January–February 2009 heatwave in south-eastern Australia. *Bulletin of the Australian Meteorological and Oceanographic Society* **22,** 13–19. Also see: <www.bom.gov.au/climate/current/statements/scs17d.pdf>

Smart RE, Alcorso C and Hornsby DA (1980) A comparison of winegrape performance at the present limits of Australian viticultural climates – Alice Springs and Hobart. *The Australian Grapegrower and Winemaker* **28** and **30**. Winetitles, Adelaide.

Smart RE and Robinson M (1991) *Sunlight into Wine. A Handbook for Winegrape Canopy Management.* Winetitles, Adelaide.

Snyder KA, Richards JH and Donovan LA (2003) Night-time conductance in C3 and C4 species: do plants lose water at night? *Journal of Experimental Botany* **54,** 861–865. <http://jxb.oxfordjournals.org/cgi/content/abstract/54/383/861>

Thorne ET, Stevenson JF, Rost TL, Labavitch JM and Matthews MA (2006) Pierce's disease symptoms: Comparison with symptoms of water deficit and the impact of water deficits. *American Journal of Enology and Viticulture* **57,** 1–11. <www.ajevonline.org/cgi/content/abstract/57/1/1>

Webb L, Watt A, Hill T, Whiting J, Wigg F, Dunn G, Needs S, Barlow EWR (2009) Extreme heat: managing grapevine response based on vineyard observations from the 2009 heatwave across South-Eastern Australia. *Australian Viticulture* **13,** pp. 39–50. Winetitles, Adelaide.

Webb L, Dunn G and Barlow EWR (2008) 'Water allocation scenarios in key regions within the Murray Darling Basin for the 2008/09 season and beyond'. GWRDC South Australia MODULE 06. <http://waterandvine.gwrdc.com.au/index.php?id=15>

White MA, Diffenbaugh NS, Jones GV, Pal JS and Giorgi F (2006) Extreme heat reduces and shifts United States premium wine production in the 21st Century. *Proceedings of the National Academy of Sciences* **103,** 11217–11222.

Zsofi Z, Varadi G, Balo B, Marschall M, Nagy Z and Dulai S (2009) Heat acclimation of grapevine leaf photosynthesis: mezo- and macroclimatic aspects. *Functional Plant Biology* **36,** 310–322.

12. GETTING ON TARGET: ENERGY AND WATER EFFICIENCY IN WESTERN AUSTRALIA'S HOUSING

Carolyn Hofmeester and Brad Pettitt

Abstract

Climate change is creating greater urgency for housing design and construction that will minimise energy and water consumption to help mitigate greenhouse gas emissions and adapt to reduced rainfall and drought conditions, particularly in southern Australia.

This chapter examines government regulations that are in place to address the energy and water efficiency of residential houses. Of particular interest is the appropriateness of Five Star Plus standards in Western Australia (WA). The findings suggest that Five Star Plus as a minimum standards approach is a necessary but inadequate step forward for sustainable housing in WA. Five Star Plus helps to eliminate worst practice, but a target-based system such as the NSW Building Sustainability Index may also be required to make substantial inroads in reducing greenhouse gas emissions, minimising potable water use and creating incentives for alternative technologies. The implications of emissions reductions through the Carbon Pollution Reduction Scheme (CPRS) for housing energy efficiency are also raised as well as steps required for more greenhouse responsive and sustainable housing.

Introduction

Western Australia, like the rest of the country, is highly vulnerable to climate change with projections of temperatures in the south-west, where most of the state's population is located, rising by up to 2°C by 2030 (summer and winter) and up to 6.5°C (summer) and 5.5°C (winter) by 2070. Rainfall will continue to decline and may fall to less than 60% of historic averages by 2070. Extreme events will increase with the number of days over 35°C predicted to rise to 70 per year for Perth. (Bates *et al.* 2008; CSIRO 2007)

The energy and water requirements of our built environment are significant sources of greenhouse gas emissions, energy consumption and water use. The embodied and operational energy of our buildings account for around 40% of our energy requirements, and the residential sector accounts for 11% of greenhouse gas emissions and 70% of total water consumed in our urban centres (Department of Climate Change 2008; GHD Pty Ltd 2006; Madew 2006).

At the same time, increasing the energy efficiency of our built environment could contribute 11% of greenhouse gas abatement by 2030 with no net cost to the economy. This potential would be achieved through efficiencies in water heating, lighting, room heating and ventilation in particular (Gorner *et al.* 2008). In other words, housing energy efficiency is the low hanging fruit in emissions reduction with a negative cost per tonne of carbon dioxide equivalent (CO_2-e) saved.

Despite the substantial impact and opportunity for low or no cost abatement, the energy and water efficiency of residential buildings in Western Australia (WA) remains weak. This chapter argues that this is due in large part to inadequate government policy and regulations.

Trends in energy and water use in housing

In line with national trends, residential per capita energy use in WA grew by 11% from 15.5 gigajoules (GJ) to 17.2 GJ between 1990 and 2005. This was caused by the greater use of electrical appliances and space heating and cooling which rose by 20% and 33% respectively (Australian Greenhouse Office 2007). Between 1970 and 2004, house sizes also increased by 92% while the number of people per household declined in the same period by 24% (Australian Bureau of Statistics 1998, 2007; Troy 1996).

The Western Australian household consumption of water at 277 kilolitres per year is higher than the national average. Half of Perth's residential potable water is being used mainly for watering gardens, equivalent to 1.3 million swimming pools of drinking-quality water per year (Department of the Premier and Cabinet 2006; Environmental Protection Authority 2007).

Towards sustainable housing: regulation at the national and state levels

The quest for greater energy efficiencies in housing regulations at the national level has been a slow journey. The genesis of the current national approach towards addressing climate change imperatives in the housing sector can be traced as far back as the 1992 National Strategy for Ecological Sustainable Development endorsed by the Council of Australian Governments (COAG). Among the objectives in the Strategy were a number directed at improving energy efficiency of residential buildings, promoting subdivision planning to maximise energy-efficient housing and switching to energy sources with lower greenhouse gas emissions (Council of Australian Governments 1992). Seven years later, in 1999, the Australian Greenhouse Office (AGO) undertook a scoping study for the incorporation of national mandatory minimum energy performance standards (rather than the existing voluntary standards) into the Building Code of Australia (BCA).

It took a further four years for the Australian Building Codes Board to introduce in 2003 the first mandatory standard (four stars) within the BCA which addressed the types of standards recommended by the AGO report. The energy efficiency provisions included insulation of building structure, insulation and sealing of air-conditioning ductwork and hot water piping, glazing, ventilation and infiltration. Within the first year of implementation, it became apparent that the four star standard was not stringent enough, but it took another two years for the standard to be raised nationally to five stars (in July 2005) due to the consultation period required and the need to allow for adequate transition to the higher standard (Australian Building Codes Board 2004; 2007). By international standards, however, a five star rating is still well short of what should be mandated. According to the International Comparison of Building Energy Performance Standards report (Horne *et al.* 2005, p. 4) the BCA Five Star standard is 1.8 to 2.5 stars below what is achieved in comparable international locations. For example, Perth was climate matched with Bakersfield, California where the star ratings for selected houses were in the range of seven to eight stars with a median rating of 7.5 stars. The difference between 7.5 stars and five stars would be an energy consumption saving of 74 MJ/m^2 for Perth or a 45% energy saving.

In the April 2009 COAG Communiqué, the state and federal governments have committed to upgrade the minimum standard to six stars by 2010 subject to a Regulatory Impact Statement. This, however, represents yet another incremental step that fails to capitalise on the opportunities of maximising residential building energy efficiency.

Another shortcoming associated with the current national approach based on the BCA is the lack of integration with water efficiency and other climate change adaptation measures such as improving resilience to the predicted increase in extreme events such as bushfires, coastal inundation and intense tropical cyclones.

While the use of the BCA is an important step towards creating a nationally consistent standard for the built environment, it is an inherently conservative and limited approach. This has perhaps been the factor influencing some state and local governments to develop policies and regulations that go beyond what the BCA can offer. For example, Western Australia and Victoria have supplemented the BCA with water efficiency measures such as alternative water sources and water-efficient appliances. NSW has taken a completely different approach by replacing the BCA with its own requirements under the NSW Environmental Planning and Assessment Act and has provided an important alternative to the BCA for achieving greenhouse gas abatement, energy and water efficiency. Under the NSW Building Sustainability Index (BASIX), mandatory targets are set for energy and water efficiency using a performance–incentive approach that allows designers, developers and builders to choose the combination of features to achieve the targets. The NSW approach also extends to housing renovations over $50 000. BASIX encourages innovation and alternative energy use by awarding the highest scores to features which result in low greenhouse gas emissions and potable water use.

The Wilkenfeld report (George Wilkenfeld and Associates 2007) found that improving the thermal performance of buildings (as prescribed by BCA type standards) does not necessarily result in reduced greenhouse gas emissions since emissions are dependent on the type of equipment and appliances used (e.g. gas or solar versus electric), their inherent efficiency and extent of use. The Wilkenfeld research estimated that the average energy-related emissions of new dwellings in Victoria (Class 1 houses and Class 2 apartments) were actually 6% higher than existing dwellings due to the growth in emissions from lighting and the increasing size of dwellings. The calculations showed that the preference for halogen and incandescent lighting over compact fluorescent lamps in new houses added 1180 tonnes of CO_2-e year. This negated the benefits of improved heating and cooling loads (thermal comfort) through the five star rating, which reduce emissions by approximately 880 tonnes of CO_2-e per year. The Wilkenfeld report proposed a new performance-based approach for Victoria using a greenhouse benchmark. The approach would combine increasing the thermal performance to a seven star rating with an assessment tool such as the NSW BASIX to achieve emissions reductions target.

Western Australian case study: standards versus targets

In Western Australia, the approach to housing energy efficiency has been to adopt the national energy efficiency standards established by the BCA. The initial standards were designed to achieve three to four star rating and were adopted in WA in July 2003. Although enhanced provisions under the BCA were introduced in 2006 to achieve a five star standard, WA elected to use the 12 month transitional period to delay mandatory implementation until May 2007. To place the star ratings in the context for Western Australia, the four star standard specifications in Perth would result in 118 MJ/m^2 energy consumption per house per year, whereas the five star standard increased the energy efficiency by a further 25% to 89 MJ/m^2 energy consumption for each house per year. Moving from five star to a seven star standard would increase energy efficiency by a further 42%.

In 2006, the WA Department for Planning and Infrastructure (DPI) commissioned an independent study into the cost–benefits of using the NSW BASIX system in Western Australia as an alternative to the BCA as the regulatory tool to achieve more sustainable housing.

The study conducted by the consulting firm Pracsys showed that setting a 40% energy efficiency target in Western Australia would reduce greenhouse gas emissions for each new four bedroom dwelling by 3240 kg CO_2-e per year or by 875 kg CO_2-e per person per year. Over a 20-year period, the savings in emissions would be 6 million tonnes of CO_2-e for the state, equivalent to $915 million reduction in energy costs to the public. Water savings were estimated to be in the order of 25 to 45 kilolitres per house per year, and over a 20-year period would result in 236 gigalitres or $275 million savings for the state (Pracsys 2006). A community survey commissioned by the Department of Planning and Infrastructure (DPI) found that 94% of homebuyers supported mandatory sustainability measures and 89% were willing to pay at least $2000 extra for the measures (Synovate 2006).

In May 2007, however, with the release of the state's Climate Change Action Statement, the proposed implementation of BASIX was abandoned in favour of continuing with the BCA standards and supplementing these with mandated water-efficiency requirements, as was done in Victoria in 2005. This new regulatory framework, which took effect on 1 September 2007, is the Five Star Plus comprising two codes – the Energy Use in House Code and the Water Use in House Code. The Energy Use Code has the BCA Five Star Standard associated with thermal comfort of the building and additional requirements for energy-efficient water heating, namely solar gas or heat pump hot water system. The Water Use Code would be satisfied by the use of tap fittings that are at least four star rated, showerheads that are at least three star rated and dual flush toilets that are four star rated. All outdoor swimming pools and spas are required to have covers that reduce evaporation and all internal hot water pipes must be insulated with outlets no more than 20 metres from the heating source or two litres of internal volume (Department of Housing and Works 2009).

To help make sense of the consequences of implementing Five Star Plus as opposed to a system like BASIX in WA, it is instructive to look at the results of modelling done by the DPI on a number of houses being constructed in WA. This modelling was undertaken by using the design specifications provided by a number of builders to make an assessment using the BASIX tool (on licence from NSW) and comparing this to the generic requirements under Five Star Plus for the same house size. The results for the sample of nine houses are provided in Table 12.1. The results indicate that firstly, the Five Star Plus requirements for energy (thermal comfort plus use of instant gas hot water system) generates less than 20% energy savings (based on average household greenhouse gas emissions), and even with a thermal comfort rating between six and eight stars, the energy savings do not rise above 23%. The average energy saving for Five Star Plus is 15%. This is consistent with the Wilkenfeld study which showed that the five star requirements in Victoria (heating and cooling loads and hot water heating) had reduced greenhouse gas emissions by only 16%. The DPI modeling using BASIX has shown that energy savings are more evident (averaging 28%) if other factors such as natural lighting, more efficient artificial lighting and alternative energy sources are taken into account.

Not surprisingly, the greenhouse gas reductions under Five Star Plus would also be significantly less than the savings of 4860 kg CO_2-e per year (for a four bedroom detached house) envisaged if the houses had been required to comply with the requirements under WA BASIX. Table 12.1 shows that the average savings using the requirements under Five Star Plus is 921 CO_2-e kg per household per year compared to average savings of 1799 CO_2-e kg per

Table 12.1   Analysis of efficiency outcomes for nine house designs in WA using BASIX and Five Star Plus

HOUSE	Star rating	WA BASIX Greenhouse gas savings (kg)	WA BASIX % savings	FIVE STAR PLUS Greenhouse gas savings (kg)	FIVE STAR PLUS % Savings
A	5	1515	23	605	9
B	5	1469	23	878	14
C	5	1183	18	397	6
D	5	1365	21	540	8
E	5	1456	22	540	13
F	5	1105	17	1203	19
G	6	2470	38	1430	22
H	7.5	2700	45	1200	20
I	8	2925	45	1495	23
Average	5.7	1799	28	921	15

Data source: WA Department of Planning and Infrastructure, pers. comm.

household per year using the WA BASIX criteria. It is evident that only three of the designs would have met the WA BASIX target of at least 35% emissions reduction. To achieve a similar result as BASIX in reducing greenhouse gas emissions, Five Star Plus would require additional mandated features such as substituting refrigerated air conditioning with evaporative cooling and/or ceiling fans, five star gas central heating, gas cooking and energy-efficient lighting fixtures.

Unfortunately, the opportunity to maximise water efficiency in WA households has been further eroded. It was initially proposed under Stage 2 of Five Star Plus that the Water Use Code be extended to include alternative internal water supply connections for non-potable uses (toilets and washing machines) and greywater diversion systems for waste water from the bathroom and laundry. In December 2008, however, the Minister for Works decided to abandon Stage 2 due to intense lobbying by the building industry against the additional cost of sustainability regulations. While the estimated increase in building costs associated with Five Star Plus is $3500, studies have shown consistently the net benefits of this investment for home owners. In the ACT, for example, house prices have been shown to be 12% higher for houses with the highest energy ratings (Department of Housing and Works 2009; Lee 2008; Moran and Novak 2009).

It is noted that the strength of the BCA is that the building design and fabric is made more energy-efficient which will last the lifetime of the house. By contrast, the target-setting approach may not result in such enduring benefits particularly if the choice of energy or water efficiency measures is not built into the fabric of the house or not 'permanently' fixed, making them vulnerable to removal by future occupants. Also, if targets are too low, then some important energy efficiency elements such as lighting may not change, as targets are achieved using other 'big ticket' items such as hot water systems and energy efficient appliances.

The introduction of the Carbon Pollution Reduction Scheme (CPRS), which is Australia's emissions trading scheme designed to place a cap on greenhouse gas emissions, adds a further dimension to this consideration given that an effective CPRS with strong emissions reductions targets should provide an adequate price signal for consumers to reduce the greenhouse gas intensity of their energy use. As part of a suite of measures to assist householders to reduce their energy cost under the CPRS, COAG has developed the National Strategy for Energy

Efficiency which includes, as highlighted earlier, an increase in energy efficiency requirements under the BCA for new residential buildings to six stars (or equivalent) to be implemented by May 2011, as well as new efficiency requirements for hot water systems and lighting.

Conclusions

Overall, the results from this study show that the BCA-based approach has the benefit of setting an absolute minimum standard that is important for the elimination of bad or worst practice. The BASIX approach, however, offers much more scope to encourage best practice and innovation and to reduce greenhouse gas emissions and potable water use, particularly if targets are set at appropriate but challenging levels. There seems to be a strong case emerging for bringing together the BCA standard and a target approach as suggested in the Wilkenfeld report.

The findings outlined in this chapter suggest that the combination of strong emissions reduction targets (which may eventually be delivered through the CPRS) plus high building standards relating to the energy and water efficiency of houses would provide greenhouse-positive and sustainable outcomes for the residential built environment. The additional national energy efficiency strategies proposed by COAG, however, are not sufficient to maximise the benefits unless the building fabric standard is raised to at least seven stars.

Improving the energy efficiency of new houses is a necessary but not sufficient step forward in lowering Australia's greenhouse gas emissions. This needs to be complemented by a range of other policy instruments that will improve the sustainability of our housing including the introduction of mandatory disclosure of the energy rating, greenhouse emissions and water efficiency at point of sale or lease (as is the case in the ACT), along with a massive program to sustainably retrofit Australia's existing housing stock. More energy- and water-efficient housing can be a low to negative cost part of the solution for reducing Australia's carbon footprint, but it is going to require a stronger regulatory approach from the governments around Australia to make sure this low hanging fruit is not left to rot on the tree.

References

Australian Building Codes Board (2004) 'Submission to the Productivity Commission Inquiry into Energy Efficiency'. Canberra.

Australian Building Codes Board (2007) 'BCA Housing Energy Measures'. ABCB. <www.abcb.gov.au>

Australian Bureau of Statistics (1998) 'Australian Social Trends, 1998 Cat. 4102.0'. Commonwealth of Australia.

Australian Bureau of Statistics (2007) 'Australian Social Trends, 2007 Cat. 4102.0'. Commonwealth of Australia.

Australian Greenhouse Office (2007) 'State and Territory Greenhouse Gas Inventories 2005'. Australia's National Greenhouse Accounts. Commonwealth of Australia, Canberra.

Bates BC, Hope P, Ryan B, Smith I and Charles S (2008) Key findings from the Indian Ocean Climate Initiative and their impact on policy development in Australia. *Climate Change* **89**, 339–354.

Council of Australian Governments (1992) 'National Strategy for Ecologically Sustainable Development'. Commonwealth of Australia. <www.environment.gov.au/esd/national/nsesd/strategy/index.html>

CSIRO (2007) 'Climate Change in Australia. Technical Report Appendix B: Cities summaries'. CSIRO, Melbourne. <www.climatechangeinaustralia.gov.au>

Department of Climate Change (2008) 'Australia's National Greenhouse Accounts. National Inventory by Economic Sector 2006'. Department of Climate Change, Canberra.

Department of Housing and Works (2009) 'Five Star Plus'. Perth, Western Australia. <www.fivestarplus.wa.gov.au>

Department of the Premier and Cabinet (2006) 'Draft State Water Plan'. Government of Western Australia, Perth.

Environmental Protection Authority (2007) 'State of the Environment Report: Western Australia 2007'. Department of Environment and Conservation, Perth, Western Australia.

George Wilkenfeld and Associates (2007) 'Options to Reduce Greenhouse Emissions from New Homes in Victoria Through the Building Approval Process'. Department of Sustainability and Environment, Victoria, Australia.

GHD Pty Ltd (2006) 'Scoping Study to Investigate Measures for Improving the Water Efficiency of Buildings'. Australian Greenhouse Office, Commonwealth of Australia.

Gorner S, Lewis A, Downey L, Slezak J, Michael J and Wonhas A (2008) 'An Australian Cost Curve for Greenhouse Gas Reduction'. McKinsey & Co, Sydney. <www.mckinsey.com/locations/australia_newzealand/knowledge/pdf/1802_carbon.pdf>.

Horne R, Hayles C, Hes D, Jensen C, Opray L, Wakefield R and Wasiluk K (2005) 'International Comparison of Building Energy Performance Standards'. RMIT Centre for Design, Melbourne, Victoria.

Lee T (2008) *Gas Guzzling Houses Lead ACT Price Decline*. ABSA Association of Building Sustainability Assessors. Surry Hills, NSW.

Madew R (2006) 'The Dollars and Sense of Green Buildings 2006'. Green Building Council of Australia.

Moran A, Novak J (2009) 'The great lock out: the impact of housing and land regulations in Western Australia'. Institute of Public Affairs <www.ipa.org.au/library/publication/1238734419_document_moran_novak-wahousing.pdf>.

Pracsys (2006) 'WA Basix: Triple Bottom Line Cost-Benefit Analysis'. Report prepared for the Department for Planning and Infrastructure. Department for Planning and Infrastructure, Perth.

Synovate (2006) 'Sustainable Housing Consumer Research Report'. Department for Planning and Infrastructure, Perth. <www.dpi.wa.gov.au>

Troy P (1996) *The Perils of Urban Consolidation*. The Federation Press, Sydney.

13. SUSTAINABLE ENERGY AS THE PRIMARY TOOL TO AMELIORATE CLIMATE CHANGE

Ray Wills

Abstract

Climate change induced by global warming will change the distribution and abundance of a huge range of species, and affect all primary industries reliant on the natural environment including agriculture, fishing, forestry, and ecotourism, and a raft of other secondary and tertiary industries that contribute to the welfare of Australian communities and support the economy. In this climate, we must use all means available to reduce and offset emissions, including strong renewable energy and energy efficiency targets. The challenge of climate change must be the catalyst for changing the way we think about and plan infrastructure, changing the way we use energy, and in using energy sustainably, future-proof our economy.

The climate debate

One role of science is to establish what (and if) relationships exist between specific events.

An enormous amount of scientific research has established that during the Earth's 4.5 billion-year history, the climate has varied and changed on a wide range of time scales, due to natural causes and without human activities having a role in climate change.

While the Earth's climate is dynamic and climate change is normal and continuous, the role of humans in causing the enhanced greenhouse effect is now theoretically and empirically well established.

Almost all scientific opinion on climate change, reported by the UN Intergovernmental Panel on Climate Change (IPCC 2007) and endorsed by the national science academies of G8 nations, as well as those of Australia, Brazil, China and India (Anon 2001), concludes that global warming is attributable to human activities.

The claim often repeated in the media that global warming is a conspiracy of climate scientists to create evidence of human-induced climate change (e.g. see Monbiot 2006; Ward 2009) lacks any substantiated evidence. The scientific community is, through the science-based process of peer review, actually good at pointing out fraud – even that which occasionally manages its way into the peer-reviewed literature.

There is a trail of evidence, however, that leads one to suspect there may be a conspiracy amongst those discounting the facts, the science and the theory. The only evidence of vested interest points to those denying the scientifically established facts supporting the theory of the anthropogenic greenhouse effect, and some evidence of key interests protecting massive value in fossil fuel resources (Plitz and Blaylock 2008; Goldenberg and Carrington 2009; Ward 2009).

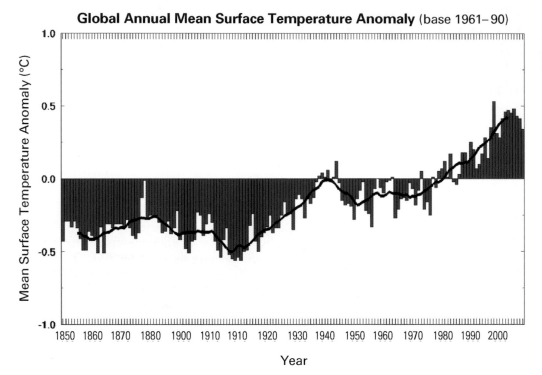

Figure 13.1: Global annual mean surface temperature anomaly (relative to 1961–1990). (Source: Australian Bureau of Meteorology)

Global temperatures are rising (see Figure 13.1). It is a fact (Hansen *et al.* 2006; IPCC 2007; Hansen 2009).

As uncertainty clings to the world's financial markets, economic and business analysts less familiar with the climate change science may find it challenging that science – with a high level of certainty – can tell us of the havoc global warming will have on our planet and by deduction on global economies, and the urgency with which we need to respond for both the world's environment and our own wellbeing.

There is one certainty in fiscal markets; that at some point that the bear will hibernate and growth will return. To be sure, the ups and downs in the market from day-to-day, week-to-week, and month-to-month cannot be forecast. But almost all economic analysts agree that the longer trends over years and decades are more regular and that growth will return and markets will recover.

Similarly, in the short-term, science cannot forecast how quickly climate changes will occur in the next months and years, and temperature and other weather changes in any given place will behave much like the stock market, especially as climate belts shift and the world's climate systems adjust to the pressures that a build-up of greenhouse gas emissions is causing. But as scientists, we are greatly concerned that the actions (and inactions) of governments and contemporary reports in the media do not appear to match the extraordinarily high level of agreement in the scientific community on the cause of human-induced climate change and the high level of risk that will continue to grow if governments fail to act.

As a consequence, the general public are not well informed enough nor yet empowered to assess the risks posed by global warming, or to provide guidance to their elected officials on appropriate responses.

In the evolutionary history of the human species, the pursuit of science is a relatively new discipline. Early cultures relied on a mix of superstition and traditional wisdom to explain the world around them. The practice of science, and the delivery of objective factual analysis of events around us, provides a fundamental challenge to traditional approaches. Nevertheless, traditional knowledge remains relevant and must be appraised.

In south-western Western Australia, the Noongar, the indigenous peoples of the land, developed an understanding of the place in which they lived. In so doing they developed a seasonal calendar divided into six (Anon. 2009). The six distinct seasons reflected the knowledge of generations and corresponded with moving to different habitats to support feeding patterns based on seasonal food. Such knowledge of the climate was not understood by European settlers of south-western Australia, who overlaid an imported system of seasons that did not acknowledge the realities of the climate of the new lands they had moved to, nor the inherent weaknesses that were exposed in attempting to dominate an environment rather than work with it sustainably.

In a similar way, throughout the 20th century, economic theory relegated most elements of the environment to being merely externalities. Such an approach has meant that by the start of the 21st century, much of our society has become divorced from the need to live in context within the environment, and has delivered unsustainable practices and a range of ills affecting the health of global ecosystems and creating the problem of global warming-induced climate change. In the later part of the 20th century, with resistance from defensive industries that had a hand in polluting the environment, regulatory changes came about that started to internalise the environmental degradation that was being caused through economic activity and previously considered 'externalities'. Water pollution and pesticides, air pollution of many forms and ecosystem degradation and the loss of ecosystem services are all in the process of being internalised; absorbed as an inevitable cost of economic activity.

Greenhouse gas emissions are the latest inclusion, and their inclusion is not a part of some green socialist agenda, but a logical requirement of creating a more sustainable approach to economic activity.

Climate change – Western Australian examples

The 2007 report from the IPCC was hailed by media reports as the 'bleakest yet' (e.g. Mason 2007). It described the increased certainty of dangerous climate change and underscored the need for increased urgency of action on global warming created by human activity.

Forecasts from the IPCC suggest global temperatures are likely to rise in the range of 1.8°C to 4.0°C this century (IPCC 2007). A warming of 1.0°C in the mid-latitudes is sufficient to move climate belts about 150 km south (UNFCC 2000), and thus a regional change of temperature of 2°C – the bottom end of the IPCC range for warming – is likely to have a catastrophic impact on many places, including Western Australia.

A 300 km shift in climatic bands will change the distribution and abundance of a huge range of species, and will affect agriculture, forestry, tourism and a range of other economic activities that contribute to the welfare of Western Australian communities (Figure 13.2).

Western Australia was arguably the first western economy with measurable economic impacts of climate change driven by global warming. The following example is just one:

- South-west Western Australia has already suffered a 20% decline in rainfall in the last 30 years – effects on runoff are potentially serious as evidenced by a 50% drop in water supply to the Perth reservoirs – and a further 20% reduction predicted that may have

The meaning of "small" warming

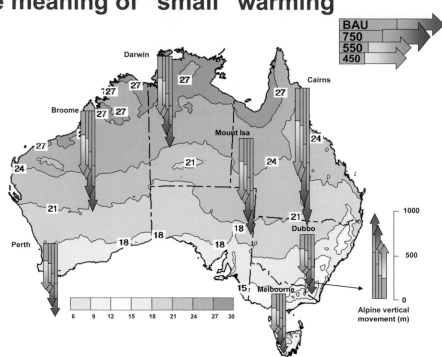

Figure 13.2: Approximate meridional and vertical movements of temperature zones by 2100 under four emissions scenarios. (Source: Pearman 2008)

started at the end of the 1990s (Hope and Foster 2005). In 2006, the Western Australian Water Corporation estimated lost income at $1 billion in WA through water restrictions and additions to infrastructure.

Climate is also a key determinant of agriculture and changes in climate will affect both crops and livestock. A number of expected changes to the agricultural industry are outlined below.

Rising temperatures will cause a shift in budburst, shorter growing seasons, earlier harvest dates and lower crop quality (UNFCC 2000). Wheat growing areas in south-west Western Australia have already been seriously affected (Kingwell 2006), and based on the output of climate change models, the northern wheatbelt is likely to disappear while production in the remainder of the wheatbelt is likely to be greatly reduced. The dairy and beef cattle industry will face decreased pasture production. Honey production in Western Australia – some of the highest production rates in the world (Wills *et al.* 1990) – is also likely to be seriously affected. Climate is a key influence in grape selection (e.g. Coomb 1987). Western Australia currently produces less than 5% of all Australian wine, but produces about 25% of the wine in the super-premium and ultra-premium categories. Shifting rainfall patterns and drier conditions will change the way vineyards operate and will reduce the wine crop (Webb *et al.* 2007).

Global warming will also affect natural ecosystems. Drying of the south coast in WA, areas with temperature increases greater than 2°C combined with a decline in rainfall consistently below 400 mm, will lead to the loss of many species of *Proteaceae* in WA's south-west, including the iconic banksia (Pouliquen-Young and Newman 1999; Pittock 2003; Lamont *et al.* 1995). In addition, the animals that rely on those species for food and shelter, including the endemic honey possum, the only mammal in the world that exists on a 100% nectar diet, are likely to die out.

Sea levels have risen in Western Australia by 18.5 cm in the last 100 years, with IPCC predictions that this will very probably triple (more than 48 cm) in the next hundred years. The potential for a one metre sea level rise by the end of this century is well within the bounds of CSIRO predictions (Church 2008; Domingues 2008). With those sorts of rises, much of the low-lying areas, including parts of Perth, Fremantle, Mandurah and Busselton/Margaret River, are under threat. Submerged fringing reefs, currently a barrier protecting parts of Perth's coastline, will be further submerged offering less protection and allowing bigger waves to reach previously sheltered beaches.

The Indian Ocean has warmed an average 0.6°C since 1960 (Levitus *et al.* 2009). Only another 0.4°C is needed for widespread and intense coral bleaching. Fortunately, strong winds during the summer results in coastal upwelling off Ningaloo Reef and the upwelling can lower the sea surface temperature by up to 3°C. Until now, this appears to have protected Ningaloo Reef from regional ocean warming (Charitha Pattiaratchi, pers. comm.).

Sustainability in a business response to climate change

One of the challenges in dealing with sustainability is communicating ideas in a way that all will understand. In the 21st century, we cannot allow our economy to play the 'externality card'. Greater understanding of complex systems has permitted the realisation that discounting factors in an economy as 'externalities' is fraught with introducing costs and transferring risk to the community. Ecological theory is penetrating experimental economics, and the lessons from competition theory and ecosystem diversity appear to apply equally to markets and trading systems. Decline in ecosystems is associated with the loss of diversity, with the establishment of counterproductive weed species. Similarly, economists agree a lack of diversity in markets can affect the economy negatively.

In 1987 the Bruntland Commission established principles for sustainable development with the following key attributes:

- deal transparently and systemically with risk, uncertainty and irreversibility;
- adopt the principle of continuous improvement;
- accept the need for good governance;
- commit to best practice;
- support no net loss of human capital or natural capital;
- ensure inter-generational equity, and
- integrate environmental, social, human and economic goals in policies and activities.

Traditionally these principles have been presented in the reverse order, and often those working on projects run the risk of treating them as a list of criteria that should be met only if possible, but none are mandatory.

Thus, a sustainability assessment may consider the economic matters and come to the conclusion that a specific project will make money, so the economic criteria are met. Similarly, the social component might consider delivery of, for example, positive outcomes for indigenous employment and the delivery of community facilities so it is concluded that the 'social sustainability' criterion has been met. Finally, in the consideration of the environmental impact of the project, researchers may find that perhaps there is a turtle nesting on a beach, and the project

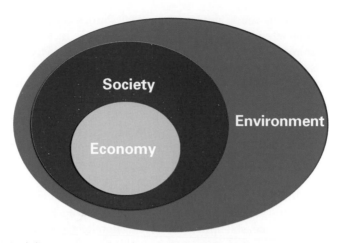

Figure 13.3: Sustainability as a nest of circles in which environment encompasses society and society encompasses the economy.

proponent might quickly conclude that 'the environment got in the way'. A fair consideration of the project should have in fact realised that, in this circumstance, the overall sustainability test for the project failed.

Sustainability cannot be understood in a list-based approach. Any approach to sustainability must be multi-dimensional. Another way this has been portrayed is a nest of circles in which environment encompasses society and society encompasses the economy (see Figure 13.3).

Sustainability can only be considered as an entity, and one element of sustainability cannot be removed from another. Sustainability cannot be compartmentalised – it is an integrated unit like a complex ecosystem, or a complex market.

The only long-term commodity is innovation

With every great crisis there is also great opportunity; clichéd but true. Rebuilding long-neglected infrastructure and investing in research, new technologies and sustainable energy are part of what defines economic opportunity in the 21st century.

The challenge today is to embed green economic policy into national economies and create markets that are able to value sustainable outcomes.

One example is provided in a 2008 UN Report, 'Green Jobs: Towards Decent Work in a Sustainable, Low-Carbon World'. The Global Green New Deal, part of the UNEP Green Economy initiative, has called for one-third of the world's $2.5 trillion-worth of planned stimulus packages to be invested in greening the world economy (Barbier 2008).

UNEP figures show that in recent years 2.3 million people have found new jobs in the renewable energy sector alone. There is great potential for job growth in the sector, with combined employment in wind and solar power expected to rise to 8.4 million by 2030.

Investments in improved energy efficiency in buildings could generate an additional 3.5 million green jobs in Europe and the United States alone (UNEP 2008).

But with the current global squeeze on credit, politicians and policy makers need to work out how to ease the effects of a shortage of credit on capital-intensive renewable energy projects.

Governments around the globe have committed to expend public funds to shore up poor decisions in investment markets. Governments must ensure investments in public monies are tied to actions that reduce greenhouse gas emissions, and businesses must consider the investment of their own funds for the same purpose.

Non-renewable resources will place pressure on markets with price variability, but renewable energy will continue to shine on us, wash up on our shores, and blow past us without additional price pressure from the market.

Renewables for regions

Renewable energy generation is generally more labour-intensive, and more broadly distributed across regions (for example, see Kammen *et al.* 2004). With a better employment factor, renewable energy projects can lead to growth of local communities in rural WA. Establishment of renewable energy generation projects will bolster a broad range of skills, particularly in agricultural regions. Biomass sources – either biomass for electricity generation or feed stocks for bioethanol and biodiesel production – in particular will draw on and build the skills already available in the regions.

An Australian Bureau of Agricultural and Resource Economics (ABARE) report (Gunasekera *et al.* 2007) suggesting a massive decline in farm production and agricultural export earnings in coming decades unless we can halt climate change or adapt to it, underscores the need to strengthen rural communities to help with the battle against climate change.

As we plan for such an outcome, the economic analysis needs to be lifecycle-based. Sustainability principles must drive cost–benefit analysis with marginal abatement curves dealing with cradle-to-cradle issues (beyond cradle to grave, an approach that innately assumes waste and disposal are acceptable outputs) and considering all values that will be delivered not just to the economy, but to the community as a whole. Poor designs that are wasteful of capital (including human and natural capital) or materials will cost money and be unsustainable.

Responding to climate change will diversify our industry base. It will create new businesses that take advantage of new opportunities, and the result will be a more sustainable economy.

The key is ensuring a diversity of energy supply for electricity production. Over-reliance on a single source of supply is not wise. If we had moved faster in the past five years to commission a variety of new renewable energy projects across Western Australia, the 2009 gas outage caused by the explosion and fire in gas export pipelines that deliver gas from Varanus Island to the domestic gas market, and restriction in supply would undoubtedly have been of less concern.

It is time to give serious commitment to develop renewable energy resources, to establish real targets for sustainable energy – that is both energy efficiency and renewable energy – and to set strong, market-based financial signals to stimulate commercial investment.

This will require multiple measures across all available technologies and using all available energy efficiency measures. Market competition is important in the long term, but in the short term new technologies will need support to make it to the commercial phase. The urgency of dealing with climate change is that these cannot be left to natural market processes to grow to that point, but significant measures must be in place to ensure that the potential of new technologies is reached as fast as possible.

To achieve sustainability, any economic analysis must fully assess all benefits to the community. Governments can regulate and ensure any growth in generation capacity comes only from a combination of renewable sources and efficiency gains through co-generation technologies in thermal generation, while simultaneously insisting on the retirement of the most inefficient, highest emissions power stations within a short time frame, to be replaced with lower emissions plants with much higher thermal efficiency.

References

Anon (The Royal Society 2001) *The Science of Climate Change*, <http://royalsociety.org/Joint-science-academies-statement-Global-response-to-climate-change>

Anon (2009) *Noongar*, <http://en.wikipedia.org/wiki/Noongar>

Barbier EB (2008) 'A Global Green New Deal'. Report prepared for the Economics and Trade Branch, Division of Technology, Industry and Economics, United Nations Environment Programme. UN, New York.

Church JA, White NJ, Aarup T, Wilson WS, Woodworth PL, Domingues CM, Hunter JR and Lambeck K (2008) Understanding global sea levels: past, present and future. *Sustainability Science* **3**, 9–22.

Coombe BG (1987) Influence of temperature on composition and quality of grapes. *Acta Horticulturae* **206**, 23–35.

Domingues CM , Church JA, White NJ, Gleckler PJ, Wijffels SE, Barker PM and Dunn JR (2008) Improved estimates of upper-ocean warming and multi-decadal sea-level rise. *Nature* **453**, 1090–1094, doi:10.1038/nature07080.

Goldenberg S and Carrington D (2009) Revealed: the secret evidence of global warming Bush tried to hide. Sunday 26 July 2009 *The Observer*, London.

Gunasekera D, Kim Y, Tulloh C and Ford M (2007) 'Climate Change impacts on Australian agriculture'. Australian Bureau of Agricultural and Resource Economics (ABARE) report <www.abareconomics.com/publications_html/ac/ac_07/a1_dec.pdf>

Hansen JE (2009) 'GISS Surface Temperature Analysis Global Temperature Trends: 2008 Annual Summation'. <http://data.giss.nasa.gov/gistemp/2008/>

Hansen JE, Sato Mki, Ruedy R, Lo K, Lea DW, Medina-Elizade M (2006) Global temperature change. *Proceedings of the National Academy of Sciences* **103**, 14288–14293.

Hope P and Foster I (2005) 'How our rainfall has changed – the south-west'. Indian Ocean Climate Initiative (IOCI) Perth, Western Australia <www.ioci.org.au/pdf/IOCIclimatenotes_5.pdf>

IPCC (2007) Climate Change 2007: The Physical Science Basis. Contribution of Working Group I to the Fourth Assessment Report of the Intergovernmental Panel on Climate Change. Cambridge University Press, Cambridge, UK.

Kammen DM, Kapadia K and Fripp M (2004) 'Putting Renewables to Work: How Many Jobs Can the Clean Energy Industry Generate?' Report of the Renewable and Appropriate Energy Laboratory. University of California, Berkeley.

Kingwell R (2006) Climate change in Australia: agricultural impacts and adaptation. *Australasian Agribusiness Review* **14**, Paper 1, p. 30.

Lamont BB, Wills RT and Witkowski ETF (1995) Threats to the conservation of southwestern Australian Proteaceae. *Acta Horticultura* **387**, 9–18.

Levitus S, Antonov JI, Boyer TP and Stephens C (2009) Warming of the world ocean. *Science* 24 March 2000 **287**, 2225–2229.

Mason J (2007) Climate predictions 'bleakest yet.' 6 April 2007 *The Australian*. <www.theaustralian.news.com.au/story/0,20867,21515206-1702,00.html>

Monbiot G (2006) *Heat: How to Stop the Planet Burning*. Allen Lane, London.

Pearman G (2008) Climate change risk in Australia under alternative emissions futures. In: *Climate change impacts and risks modelling of the macroeconomic, sectoral and distributional implications of long term greenhouse-gas emissions reduction in Australia*. Graeme Pearman Consulting Pty Ltd. <www.treasury.gov.au/lowpollutionfuture/consultants_report/downloads/Risk_in_Australia_under_alternative_emissions_futures.pdf>

Pittock B (Ed.) (2003) *Climate Change: An Australian Guide to the Science and Potential Impacts*. Australian Greenhouse Office, Canberra.

Plitz R and Blaylock D (2008) *New Climate Report Counters Bush Administration's Record of Denial, Disinformation, Cover-Up, and Delay*. 29 May 2008 <www.climatesciencewatch.org/index.php/csw/details/new_government_scientific_assessment>

Pouliquen-Young O and Newman P (1999) 'The Implications of Climate Change for Land-Based Nature Conservation Strategies. Final Report 96/1306'. Australian Greenhouse Office,

Environment Australia, Canberra, and Institute for Sustainability and Technology Policy, Murdoch University, Perth, Australia,

UNEP (2008) 'Green Jobs: Towards Decent Work in a Sustainable, Low-Carbon World'. UNEP/ILO/IOE/ITUC, September 2008. UN. New York. <www.unep.org/labour_environment/PDFs/Greenjobs/UNEP-Green-Jobs-Report.pdf>

UNFCCC (2000) *Climate Change Information Sheet 10* United Nations Framework Convention on Climate Change Bonn, Germany. <http://unfccc.int/cop3/fccc/climate/fact10.htm>

Ward R (2009) *Why ExxonMobil must be taken to task over climate denial funding.* The Guardian, Wednesday 1 July 2009. <www.guardian.co.uk/environment/cif-green/2009/jul/01/bob-ward-exxon-mobil-climate>

Webb L, Whetton P and Barlow E (2007) Modelled impact of future climate change on the phenology of winegrapes in Australia. *Australian Journal of Grape and Wine Research.* **13**, 165–175.

Wills RT, Lyons MN and Bell DT (1990) The European honey bee in Western Australian kwongan – foraging preferences and implications for management. In *Australian Ecosystems: 200 years of Utilization, Degradation and Reconstruction.* (Proceedings of the Ecological Society of Australia, Vol. 16) (Eds D. A. Saunders, A. J. M. Hopkins, R. A. How) pp. 167-176. Surrey Beatty & Sons Pty Ltd, Chipping Norton.

14. A NATIONAL ENERGY EFFICIENCY PROGRAM FOR LOW-INCOME HOUSEHOLDS: RESPONDING EQUITABLY TO CLIMATE CHANGE

Damian Sullivan and Josie Lee

Abstract

Household energy efficiency can play an important role in reducing the regressive impact of energy price rises including increases resulting from an emissions trading scheme. Household energy efficiency can also provide many other benefits which include emission reductions, climate change adaptation benefits through reduced exposure to heatwaves, improvements in health and wellbeing, and employment benefits. While the Australian government has focused its Carbon Pollution Reduction Scheme household compensation package on direct payments, there remains a role for a comprehensive energy efficiency program for low-income households. The KPMG (2008) and Brotherhood of St Laurence proposal to retrofit 3.5 million low-income households illustrates the level of savings that could be achieved. A funded proposal is essential because there are significant barriers and market failures to the uptake of household energy efficiency measures in low-income households. To address these barriers and other market failures, state and federal governments should develop programs which leverage spending in existing energy efficiency programs, particularly the federal insulation program, and provide comprehensive home energy efficiency programs for low-income households. To ensure flexibility and to maximise energy efficiency and carbon savings, home visits should be a key element of these programs.

Introduction

> The price imposed by an emissions trading scheme is not intended to result in large, arbitrary transfers of wealth, especially regressive changes in income distribution. There is a clear role for government in ensuring distributive efficiency and addressing the social welfare implications of climate change mitigation policy on those people who are most affected by an emissions price and least able to respond …
>
> Garnaut Climate Change Review, 2008a, p. 389

The need for Australia to take timely and ambitious steps to mitigate greenhouse gas emissions on environmental, economic and social grounds is clear (see Allen Consulting Group 2006a; Garnaut 2008a; IPCC 2007; McKinsey and Company Australia 2008; Preston and Jones 2006; Stern 2006). In the attempts to achieve this goal, it will be important not to neglect the equity implications of climate change mitigation and adaptation policies and actions.

The design and implementation of climate change policies will have an impact on three dimensions of equity: equity between nations, equity between generations and equity within nations (Garnaut 2008b). The international equity dimensions of climate change are recognised in the United Nations Framework Convention on Climate Change, particularly in the use of the phrase, 'common but differentiated responsibility', and will be essential in achieving an effective post-2012 global climate agreement for mitigation and adaptation (UNFCCC 1992). Inter-generational equity is central to debates about the weighting we place on climate change actions, including the balance between current economic costs, and predicted future economic, environmental and amenity losses resulting from climate change. In relation to climate change, equity within nations refers to the distribution of costs and benefits of policy responses to climate change and the impacts of climate change itself. This chapter focuses on equity within Australia, particularly the distributional impacts of climate change mitigation policy.

The contention of this chapter is that household energy efficiency can play an important role in reducing the regressive impact of energy price rises including increases resulting from an emissions trading scheme. Household energy efficiency can also provide many other benefits which include emission reductions, climate change adaptation benefits through reduced exposure to heatwaves, improvements in health and wellbeing, and employment benefits. While the Australian government has focused its Carbon Pollution Reduction Scheme (CPRS) household compensation package on direct payments, there remains a role for a comprehensive energy efficiency program for low-income households. A funded proposal is essential because there are significant barriers and market failures to the uptake of household energy efficiency measures in low-income households. Policies enabling low-income households to enact energy efficiency measures are examples of climate-related policy measures that have multiple benefits.

Distributional impacts of the CPRS

The Australian Government's primary climate change mitigation policy response is a national emissions trading scheme – the CPRS (DCC 2008). The merits of an Australian emissions trading scheme have been discussed extensively elsewhere (see, for example, ABARE 2006; Allen Consulting Group 2006b; Garnaut 2008a; McLennan Magasanik Associates 2008). While the CPRS may be a useful tool to drive least-cost emission reductions through fuel switching and efficiency improvements, the price increases which will follow from a price on greenhouse gas emissions will have unintended distributional consequences.

Garnaut (2008a) explains that the additional cost of carbon emissions faced by electricity generators and gas suppliers will be passed down the supply chain to distributors and retailers and finally to households through higher energy prices. Food and other essential goods and services will also rise in price due to the emissions and energy embedded in their production, as will petrol if the initial fuel tax offset is lifted. As a result, the CPRS will lead to an increase in the cost of living, particularly the price of essential services. Low-income households who spend a greater proportion of their income on basic necessities than other households will be disproportionately affected (Garnaut 2008a, NIEIR 2007). This is the case even though they consume less energy-intensive goods and services overall.

Modelling by NIEIR (2007) for the Brotherhood of St Laurence (BSL), which compared the carbon price impact on 20 different household types, highlighted the regressive impact. The most pronounced of these was apparent when comparing a working age social security dependent family who would face an additional 2.2% annual expenditure or A$571 from a A$25 per tonne carbon price, using utility-adjusted figures, with a high income tertiary educated household who would face 0.4% additional annual expenditure or A$702 from the same carbon price, using utility adjusted figures (see Table 14.1) (NIEIR 2008). Studies of cap-and-trade schemes in the United States also identified regressive impacts of carbon pricing regimes (CBO 2000). Subsequent analysis by Unkles and Stanley (2008), based on the NIEIR data, showed that households in outer suburbs and regional areas would face a higher price burden as a result of their greater reliance on car transport and increased travel distances.

More recent modelling such as that carried out by Garnaut (2008a) and by Treasury for the Department of Climate Change (DCC 2008) confirmed the regressive impact; however, they provided revised estimates of the degree. The Treasury modelling projected that with a A$25 per tonne of carbon price, the cost of living could be expected to rise by about 1.1% in 2010–11 (DCC 2008). This, on average, would equate to additional expenditure of A$4 per week for electricity and A$2 per week for gas and other household fuels (DCC 2008). A single pensioner household in the lowest income quintile (i.e. the bottom 20% of households) faces an above-average price increase of approximately 1.4% in 2010–11 (DCC 2008).

Treasury modelling may underestimate the ongoing costs to households as these costs are based on the current CPRS design, which excludes emissions from agriculture, forestry, and offsets petrol price increases in the early years of the scheme. With the inclusion of these emissions sources in the scheme in the years ahead, even at a A$25 carbon price, the cost to households is likely to rise substantially.

Garnaut (2008a, p. 388) identifies that the financial impact on low-income households will be exacerbated, in absolute terms and relative to higher income households, by 'financial constraints faced by lower income households in switching to less emissions-dependent lifestyles – such as energy efficient appliances, household retrofits and vehicle type and fuel use'. Wilkins (2008, p. 4) came to similar conclusions as Garnaut, stating that, 'for lower income households, there is likely to be a more severe impact from higher energy prices and less capacity to do something about it'.

Importantly, the CPRS is not the only pressure on energy prices. From 2000–2008, the average retail electricity prices have increased across Australia by 50% (KPMG 2008). By 2020 retail electricity prices are forecast to increase in nominal terms (i.e. including inflation) by 25–52% without the CPRS. With the CPRS, this is set to rise by 30–70% (KPMG 2008). Electricity prices are expected to rise due to a variety of factors including inflation, rising transmission and distribution upgrade costs, drought (infrastructure degradation and lower flows though hydroelectricity dams), expansion of renewable energy, upgrading to the smart grid and smart meters, and increases in peak demand (Hatfield-Dodds and Denniss 2008; KPMG 2008; NERA 2008).

Responses to the disproportionate impact

A series of studies have emerged in Australia and internationally that analyse methods to reduce the regressive impacts of a cap-and-trade system (see CBO 2000; Dinan 2009; Garnaut 2008a). A key theme is the importance of selling emissions allocations as a means of generating revenue that can be used to offset the distributional impacts of emissions trading (CBO 2000; Garnaut 2008a). Recycling revenue through the tax-transfer system is the preferred method for addressing the majority of the distributional impacts (Garnaut 2008a, p. 349). The Australian Government chose this approach to address the distributional impacts of the CPRS. The measures which are outlined in the Carbon Pollution Reduction Scheme Amendment (Household Assistance) Bill 2009 explanatory memorandum provide direct compensation through a complex series of payments and tax incentives using the existing tax-transfer system. Low-income households should receive full compensation for the additional costs from the scheme, while middle income households will receive partial compensation.[1]

Table 14.1 Australian household types – impact of carbon price at $25 and $50 per tonne

Carbon price	Carbon costs additional annual expenditure								
	A$ (2006)		% of annual expenditure		Utility-adjusted A$ (2006)		Utility-adjusted % of annual expenditure	
	$25	$50	$25	$50	$25	$50	$25	$50
Household type								
Working age social security dependent family (type 1)	$584.7	$1169.4	2.2	4.4	$571.70	$1143.30	2.2	4.3
High income tertiary educated households	$1445.5	$2890.9	1.5	2.9	$368.70	$737.40	0.4	0.7
Average	$804.7	$1609.3	1.6	3.2	$351.10	$702.30	0.7	1.4

Source: adapted from NIEIR 2007, p. 19

The role of energy efficiency

The adjustments to the tax-transfer system announced by the federal government in response to the CPRS will provide a much-needed buffer for low-income households. They do not, however, promote action that is necessarily complementary to the CPRS, nor will they address the market failures and other barriers low-income households face to introducing energy efficiency measures in their own homes. These issues are the focus of the rest of this chapter.

Energy efficiency to offset energy price increases

While direct financial compensation has been the primary focus of policy makers, attention has also been paid to the role energy efficiency can play in reducing the disproportionate impact of a carbon price and other energy price rises (ACF *et al.* 2008; Garnaut 2008a; Hatfield-Dodds and Denniss 2008; KPMG 2008). In September 2008, the BSL in partnership with KPMG released a proposal for a National Energy Efficiency Program (NEEP) to assist low-income households (KPMG 2008). The program proposed retrofitting 3.5 million low-income households – around 40% of all households – over seven years to offset the distributional impacts of the CPRS.

The BSL and KPMG proposal recommended a customised home energy audit which would be followed by energy efficiency improvements up to the value of A\$2000 (including the cost of the visit) or A\$6000 for households with high levels of needs. Home energy audits are an essential part of the NEEP as they would enable tailored solutions for different households which are more likely to yield sustainable results over time (KPMG 2008; see also Wilkins 2008). Home visits would enable the measures to be tailored to key variables which include climatic factors, dwelling type, building design and materials, tenancy conditions, cultural and behavioural factors, location and household composition. Energy audits can also play a role in behaviour change. The suite of energy efficiency improvements proposed in the NEEP include compact fluorescent light bulbs, efficient shower roses, weather proofing, curtains, ceiling insulation, efficient refrigerators and in special circumstances energy-efficient water heating or air conditioning (KPMG 2008). The estimated costs and projected savings for individual measures are outlined in Table 14.2. Please note that these measures are not an exhaustive list of the potential energy efficiency improvements.

The proposal identified total savings that could accrue to low-income high energy-usage households of between \$313–\$470 per annum, for the standard package of \$2000, at a \$20/t CO_2-e price under the CPRS, increasing proportionately with the price of carbon. For the upper-bound package, up to a value of \$6000, up to \$700 per annum can be saved by households at \$20/t CO_2-e price and again, savings rising as the price of carbon increases. These savings are locked in for the lifetime of the products retrofitted.

The total cost of the program would be approximately \$8.7 billion in net present value terms over the period 2010/11–2021/22. Assuming the \$20 CO_2-e price increases by approximately 5.45% per annum over the same period, the savings that would accrue from a national energy efficiency program to households is approximately \$14.0 billion, representing a total benefit to the community of A\$5.3 billion, in net present value terms (KPMG 2008).

Both Wilkins (2008) and Garnaut (2008a) supported the idea of an energy efficiency program to assist low-income households respond to the carbon price. Garnaut (2008) proposed a voucher scheme whereby every household is provided with a voucher for \$1000 in energy efficiency products (Garnaut 2008a).

Energy efficiency, low-income households and market failures

The BSL and KPMG proposal is based on the notion that without government assistance, large numbers of low-income households will be unable to take up energy efficiency measures

Table 14.2   Energy efficiency measures estimated costs and savings ($)

Retrofit activity	Average cost/ household (incl. installation)	Proportion of dwellings that are suitable	Savings to energy bills weekly	Savings to energy bills yearly (at $20/t C-e)
Upgrade household with CFLs	$70*	65–80%	$2.02	$105
Weather sealing retrofit	$420	~75%	$2.16	$112
High efficiency showerhead	$95	~75%	$0.79	$41
Ceiling insulation	$1530	40%	$4.05	$210
Hot water – old electric to solar	$3500	40%	$8.84	$460
Hot water – old electric to heat pump	$4000	15%	$5.62	$293
Fridge upgrade	$950	17%	$1.06	$55

Source: Adapted from KPMG 2008, p. 22

because of persistent market failures and other barriers. These barriers will mean that low-income households are unlikely to respond to a price signal that flows from the CPRS.

A lack of access to capital to invest in energy efficiency measures is a basic and fundamental barrier that will restrict many low-income households from investing in energy efficiency. Other barriers include bounded rationality such as discount rates and price, inelasticity in demand, information failures, transaction and search costs, and misplaced and split incentives (DPI 2008, pp. 25–53). Studies suggest that most households discount future savings from energy efficiency, and low-income households place higher discounts on future savings from energy efficiency measures (DPI 2008). Residential electricity demand is also relatively inelastic (NIEIR cited in DPI 2008). This is particularly the case for those low-income households who are limiting their electricity consumption to minimise cost. As a result, households will not change their energy consumption based on price alone.

The split incentive between landlord and tenant is also particularly important. Landlords who pay for (or approve) government-subsidised energy efficiency measures in their rental properties do not benefit from savings on their energy bills and experience time and effort barriers, so there is less incentive to act. While the tenants do benefit from reduced energy bills, they are unlikely to put in place measures that are not transportable, such as insulation. Tenants also do not have the power to approve changes and may be wary of approaching landlords due to concerns that the rent will be increased.

Current distribution of energy efficiency measures in low-income households

While the data available on the incidence of energy efficiency measures in low-income households and private rental properties is limited, it is consistent with our understanding of the barriers to the uptake of energy efficiency.[2] The Victorian Utility Consumption Household survey (Roy Morgan Research 2008) provides a detailed breakdown of energy efficiency measures by non-concession and concession households and by tenure (see Figure 14.1 and 14.2). Taking insulation as an example, according to the survey homeowners and buyers were 'substantially more likely to have some insulation (89%), than were either private (33%) or public (43%) renters' (Roy Morgan 2008, p. 57). Insulation rates were high for households with

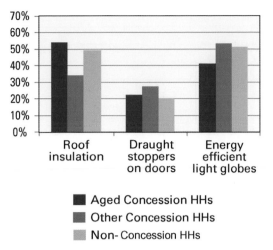

Figure 14.1: Incidence of households with full insulation by concession status. (Source: Data from Roy Morgan Research 2008, p. 58)

no concession card holders and households with aged concession households, with 74% of both household types completely insulated (Roy Morgan 2008). Other concession households, that is households with concession card holders who were not on the aged concession, had noticeably lower rates of insulation with only 51% fully insulated (Roy Morgan 2008). The higher rates in aged concession households may be explained by the high incidence of home ownership for aged pensioners, which is around 85%, with slightly lower rates for single aged pensioner households (NATSEM 2009). Other concession households were also more likely to have made low-cost energy efficiency modifications such as energy-efficient light globes or draught-stoppers than either aged concession households or non-concession households (see Figure 14.4) (Roy Morgan 2008). The information in the Roy Morgan survey is useful; however, the reliance on self-reporting for the incidence of insulation creates some limitation on the veracity of the data.

Other benefits of household energy efficiency

The BSL and KPMG proposal focused attention on the financial savings available from energy efficiency. There are, however, significant other benefits which have not been adequately quantified or realised in Australia. Household energy efficiency is likely to result in carbon savings.

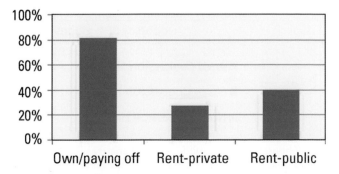

Figure 14.2: Incidence of households with full insulation by tenure. (Source: Data from Roy Morgan Research 2008, p. 58)

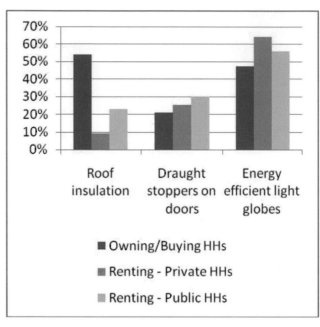

Figure 14.3: Incidence of households who made selected energy savings modifications by tenure type. (Source: Data from Roy Morgan Research 2008, p. 199)

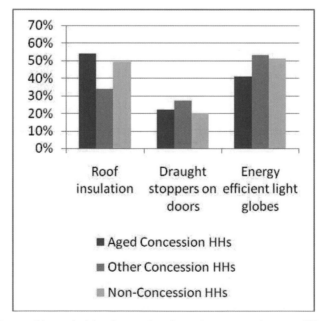

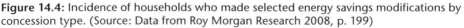

Figure 14.4: Incidence of households who made selected energy savings modifications by concession type. (Source: Data from Roy Morgan Research 2008, p. 199)

KPMG (2008) estimated that over the seven years of the NEEP program, there would be locked in lifetime CO_2-e abatement savings of approximately 45 million tonnes from the retrofit measures. A number of energy efficiency measures would also provide climate change adaptation benefits, particularly insulation and sealing cracks, which would contribute to more moderate household temperatures and reduced exposure to extreme heat events that are predicted to increase in most of Australia as a result of climate change. Individual householders in cooler climates would also be likely to benefit from improved health outcomes from warmer, drier houses (Howden-Chapman *et al.* 2007). There may also be other health benefits such as reduced stress and improvements in mental health as identified in the Warm Front evaluation in the United Kingdom (Green and Gilbertson 2008). There are also likely to be a series of economic benefits including employment stimulation. KPMG modelling identified that 40 000 direct and indirect jobs would be created if the NEEP program were introduced (KPMG 2008). Other studies have identified economic benefits from freeing up householder finances to spend in more productive areas of the economy (Roland-Holst 2008).

Existing programs

A number of existing federal and state energy efficiency programs provide low-income households with access to energy efficiency measures. They include the federally funded Home Insulation Program (HIP), the Green Loans Program and state-based energy efficiency targets. There are, however, barriers to the widespread uptake by low-income households. The federal government's redesigned A\$3.2 billion Home Insulation Program (HIP) (which was formerly two programs – the HIP program and the Low Emissions Assistance Plan for Renters), which will run for two-and-a-half years as part of the Nation Building Stimulus Plan, is the largest program. The program seeks to use competitive market pressures to drive uptake in households, which is likely to lead to high levels of uptake in areas and sectors with low access and recruitment costs, particularly capital city owner–occupiers. It may, however, disadvantage households in areas that have higher access and recruitment costs and as a result are likely to receive less attention. These include remote and rural areas, and areas with high levels of disadvantage. After very slow uptake in private rental properties, the initial insulation program was modified in September 2009 to enable private renters to receive \$1600 worth of insulation, which makes them eligible for the same benefits as owner–occupier properties. There remain considerable barriers to increasing the uptake in private rental properties. The Commonwealth has also launched a scaled-back green loans scheme which includes access to a home energy audit. The green loans are unlikely to be a priority for low-income households; however, the audit is a useful free measure if it provides an information point for households to gain access to free or low-cost efficiency measures.

There are also a variety of rebates schemes, such as the Victorian Government's Regional and Metropolitan hot water rebates. The applicability of these rebate schemes to low-income households is debatable. Wilkins (2008, p. 107) argues that rebates are often too prescriptive and assist only a limited pool of households, which are likely to be in the middle-income brackets, and are unlikely to target assistance towards low-income households.

A number of states have now introduced schemes specifically targeting low-income households. These include NSW's \$63 million Low Income Household Refit Program; Victoria's Energy and Water Taskforce and the recently announced \$2.8 million Climate Change Proofing Low Income Households Package. Victoria, NSW and South Australia also have obligations on retailers to reduce household emissions. South Australia's Residential Energy Efficiency Scheme (REES) is the only one which obliges energy providers to achieve at least one third of the savings from low-income households, and provide 13 000 energy audits for low-income households (2009–2011).

Opportunities for a comprehensive energy efficiency solution for low-income households

The provision of multiple energy efficiency programs still leaves a need for a comprehensive energy efficiency solution targeting low-income households. A comprehensive program could assist householders to participate in the multiple existing programs. It could address information barriers and highlight any significant behaviour changes required to realise energy efficiency savings. Such a program could be created by providing funding for households to undertake home energy audits that would link householders with existing schemes. A limited amount of funding would also be needed for measures not covered by the existing programs, such as sealing gaps and cracks. Such a scheme could be funded federally to ensure coverage in all states. Alternatively if federal funding is not forthcoming, individual states could expand existing programs targeting low-income households. While the current programs differ in scale, more funding to expand the existing programs is needed in all jurisdictions. In order to take advantage of the generous insulation program, such a scheme would ideally be put in place soon.

A comprehensive, affordable and targeted energy efficiency solution targeting low-income households would go some way to insulating low-income households from increased energy prices and would enable a series of unrealised benefits, including carbon emissions reduction, better health and increased employment to be realised. Such a program would be complementary to existing energy efficiency measures such as mandatory energy performance standards (MEPS) and new building standards. It would also complement moves to introduce standards for existing owner–occupier or rental properties, and/or mandatory disclosure of a dwelling's energy efficiency rating.

Policies enabling low-income households to take up energy efficiency measures are an example of climate-related policy measures that have multiple benefits. Household energy efficiency can help ensure more equitable price based climate change mitigation policies. These policies also address climate mitigation and adaptation. They should be prioritised in developing equitable responses to climate change.

Endnotes

1   The total cost of the measures in the CPRS Household Assistance Bill 2009 is A$1.64 billion in increased payments in 2011–12 and A$2.97 billion in 2012–13; with a further A$400 million in forgone revenue in 2011–12 and A$1.76 in 2012–13 from the tax offsets.
2   Department of Environment, Water, Heritage and the Arts (DEWHA 2008), provides a useful and detailed state-level analysis of residential energy usage trends. The Australian Bureau of Statistics (ABS 2008) survey provides household-level data on the numbers of households with various energy efficiency measures. However, the data are not broken down into frequency of energy efficiency measures by tenancy type or by income.

References

Allen Consulting Group (2006a) 'Deep Cuts in Greenhouse Gas Emissions – Economic, Social and Environmental Impacts for Australia'. Report to the Business Roundtable on Climate Change, Sydney. <www.businessroundtable.com.au/pdf/GHG2050_FINAL.pdf>

Allen Consulting Group (2006b) 'The Economic Impacts of a National Emissions Trading Scheme – Final Report'. Report to National Emissions Trading Taskforce, The Allen Consulting Group, Sydney. <www.ret.gov.au/Documents/mce/energy-eff/nfee/_documents/consreport_09_.pdf>

Australian Bureau of Agricultural and Resource Economics (ABARE) (2006) 'Economic Impact of Climate Change Policy: the role of technology and economic instruments'. ABARE, Canberra, <www.abareconomics.com/publications_html/climate/climate_06/cc_policy_nu.pdf>.

Australian Bureau of Statistics (ABS) (2008) 'Environmental issues: energy use and conservation Australia, March 2008'. ABS, Canberra. <www.ausstats.abs.gov.au/ausstats/subscriber.nsf/0/C7052 1268BC1B3D4CA25750E001131EF/$File/4602055001_mar%202008.pdf>

Australian Conservation Foundation (ACF), Australian Council of Social Services (ACOSS) and Choice (2008) 'Energy & Equity – Preparing households for climate change: efficiency, equity, immediacy'. Australian Conservation Foundation, Australian Council of Social Services and Choice. <www.acfonline.org.au/uploads/res/equity.pdf>

Congressional Budget Office (CBO) (2000) 'Who gains and who pays under carbon-allowance trading? The distributional effects of alternative policy designs'. Congressional Budget Office, Washington. <www.cbo.gov/ftpdocs/21xx/doc2104/carbon.pdf>

Department of Climate Change (DCC) (2008) *Carbon Pollution Reduction Scheme: Australia's low pollution future – White Paper*. DCC, Canberra. <www.climatechange.gov.au/whitepaper/report/index.html>

Department of Energy, Water, Heritage and the Arts (DEWHA) (2008) 'Energy use in the Australian residential sector 1986–2020'. DEWHA, Canberra.

Department of Environment, Water, Heritage and the Arts (2009) *Energy Efficiency Homes Package*. DEWHA, Canberra. <www.environment.gov.au/energyefficiency/>

Department of Energy, Water, Heritage and the Arts (DEWHA) (2008) 'Energy use in the Australian residential sector 1986–2020'. DEWHA, Canberra.

Department of Primary Industries (DPI) (2008) 'Proposed Victorian energy efficiency target regulations 2008 regulatory impact statement'. State of Victoria Department of Primary Industries, Melbourne. <www.dpi.vic.gov.au/DPI/dpinenergy.nsf/LinkView/D3B3CB4C58B5B203CA2574BA0 005208B4CAC723B1D538D66CA25740C000D2004/$file/Victorian%20Energy%20Efficiency%20 Target%20Scheme%20Regulatory%20Impact%20Statement.pdf>

Dinan T (2009) *The distributional consequences of a cap-and-trade program for CO_2 emissions*. Statement before the subcommittee on income security and family support committee on ways and means, United States, House of Representatives. <www.cbo.gov/ftpdocs/100xx/doc10018/03-12-ClimateChange_Testimony.1.1.shtml>

Garnaut R (2008a) *The Garnaut Climate Change Review*. Cambridge University Press, Cambridge, UK, <www.garnautreview.org.au/CA25734E0016A131/pages/garnaut-climate-change-review-final-report>.

Garnaut R (2008b) Climate Change as an Equity Issue. In: *Proceedings of the Brotherhood of St Laurence 2008 Sambell Oration*. Brotherhood of St Laurence, Melbourne. <www.bsl.org.au/pdfs/Sambell_Oration_2008_Garnaut.pdf>

Green G and Gilbertson J (2008) *Warm Front Better Health: Health Impact Evaluation of the Warm Front Scheme*. Centre for Regional, Economic and Social Research, Sheffield Hallam University, Sheffield. <www.shu.ac.uk/research/cresr/downloads/CRESR_WF_final+Nav%20(2).pdf>

Hatfield-Dodds S and R Denniss (2008) *Energy Affordability, Living Standards and Emissions Trading: Assessing the social impacts of achieving deep cuts in Australian greenhouse emissions, June 2008*. CSIRO Sustainable Ecosystems, Canberra. <www.climateinstitute.org.au/index.php?option=com_content&task=view&id=189&Itemid=40>

Howden-Chapman P, Matheson A, Crane J, Viggers H, Cunningham M, Blakely T, Cunningham C, Woodward A, Saville-Smith K, O'Dea D, Kennedy M, Baker M, Waipara N, Chapman R, and Davie G (2007) Effect of insulating existing houses on health inequality: cluster randomised study in the community. *BMJ* **334**, 460–460.

IPCC (2007) Climate Change 2007: Synthesis Report. Contribution of Working Groups I, II and III to the Fourth Assessment Report of the Intergovernmental Panel on Climate Change IPCC, Geneva, Switzerland. <www.ipcc.ch/pdf/assessment-report/ar4/syr/ar4_syr_frontmatter.pdf>

KPMG (2008) 'A National Energy Efficiency Program to assist low-income households'. KPMG, Brotherhood of St Laurence and Ecos Corporation, Melbourne. <www.bsl.org.au/pdfs/KPMG_national_energy_efficiency_program_low-income_households.pdf>

McKinsey & Company Australia (2008) 'An Australian Cost Curve for Greenhouse Gas Reduction'. McKinsey & Company Australia. <www.theaustralian.news.com.au/files/greenhouse.pdf>

McLennan Magasanik Associates (2008) 'A comparison of emission pathways and policy mixes to achieve major reductions in Australia's electricity sector greenhouse emissions'. Report to The Climate Institute, South Melbourne, <www.climateinstitute.org.au/images/reports/mmagr.pdf>.

National Centre for Social and Economic Modelling (NATSEM) (2009) 'Reform of the Australian retirement income system research report'. NATSEM, Canberra. <www.bsl.org.au/pdfs/NATSEM_BSL_Reform_of_Australian_retirement_income_system.pdf>

National Institute of Economic and Industry Research (NIEIR) (2007) 'The impact of carbon prices on Victorian and Australian households'. Report prepared for the Brotherhood of St Laurence, Melbourne. <www.bsl.org.au/pdfs/NIEIR_Impact_of_carbon_prices_on_Vic&Aust_households_final_May2007.pdf>

NERA Economic Consulting (NERA) (2008) 'Cost Benefit Analysis of Smart Metering and Direct Load Control: Overview Report for Consultation'. Report for the Ministerial Council on Energy Smart Meter Working Group, NERA Economic Consulting, Sydney. <www.nera.com/image/PUB_SmartMetering_Overview_Feb2008.pdf>

Preston B and Jones R (2006) 'Climate Change Impacts on Australia and the Benefits of Early Action to Reduce Global Greenhouse Gas Emissions'. A consultancy report for the Australian Business Roundtable on Climate Change. CSIRO. Canberra, Australian Capital Territory. <www.csiro.au/files/files/p6fy.pdf>

Quiggin J (2009) Discounting and Intergenerational Equity. In: *Climate Change and Social Justice*. (Ed. J Moss) pp. 67–81. Melbourne University Publishing, Melbourne.

Roland-Holst D (2008) 'Energy Efficiency, Innovation, and Job Creation in California'. Research Papers on Energy, Resources, and Economic Sustainability, Center for Energy, Resources and Economic Sustainability (CERES), University of California, Berkeley, California. <http://are.berkeley.edu/~dwrh/CERES_Web/Docs/UCB%20Energy%20Innovation%20and%20Job%20Creation%2010-20-08.pdf>

Roy Morgan Research (2008) 'Victorian utility consumption household survey 2007'. Department of Human Services, Melbourne. <http://dhs.vic.gov.au/__data/assets/pdf_file/0008/307358/VUCHS-2007-Final-Report.pdf>

Stern N (2006) *Economics of Climate Change – the Stern Review*. Cabinet Office, HM Treasury, London. <www.hm-treasury.gov.uk/sternreview_index.htm>

United Nations Framework Convention on Climate Change (UNFCCC) (1992) *United Nations Framework Convention on Climate Change*. UNFCCC, Bonn. <http://unfccc.int/resource/docs/convkp/conveng.pdf>

Unkles B and Stanley J (2008) 'Carbon use in poor Victorian households by local government area'. Brotherhood of St Laurence, Fitzroy. <www.bsl.org.au/pdfs/Unkles&Stanley_Carbon_use_poor_Vic_households_by_LGA.pdf>

Wilkins R (2008) 'The strategic review of Australian Government climate change programs'. Department of Finance and Deregulation, Canberra. <www.finance.gov.au/Publications/strategic-reviews/docs/Climate-Report.pdf>

15. APPLYING A CLIMATE CHANGE ADAPTATION DECISION FRAMEWORK FOR THE ADELAIDE–MT LOFTY RANGES

Douglas Bardsley and Susan Sweeney

Abstract

The challenges for regional natural resource planners and managers to adapt to projected climate change are significant across Australia. The establishment of a regional governance approach to natural resource management accentuates the pressure on local decision makers to make the correct decisions for the long-term sustainability of systems, particularly in light of the warming, drying trend projected for much of southern Australia. This chapter describes research undertaken in a partnership between the Federal and South Australian Governments, the Adelaide and Mt Lofty Ranges Natural Resource Management Board and the regional community to examine approaches to generate information to support regional adaptation decision-making. The work demonstrates that there are approaches available to regional stakeholders to help guide effective decision making for sustainable adaptation of natural resource management systems. The approaches applied here were scenario modelling, applied and participatory GIS modelling, environmental risk analysis and participatory action learning. The research has assisted the Natural Resource Management Board, local councils, key industry bodies and others to incorporate climate change risk into their strategic and investment planning.

Introduction

The challenges for regional natural resource management (NRM) to design and implement effective climate change adaptation strategies are significant across Australia. Already, recent experienced climatic conditions in the Murray–Darling Basin have, in part, led to the collapse of important water management systems in South Australia, and projections suggest ongoing drying can be expected (Potter *et al.* 2008). In another example, forecasting of the impacts of climate change on biodiversity suggest that current Australian conservation approaches are going to be fundamentally challenged in the future (Hughes 2003; Botkin *et al.* 2007). Although uncertain, the potential effects of climate change on South Australia are likely to be negative for many stakeholders. Farmers, experiencing less rainfall and greater evapotranspiration, will need to manage pastures, livestock and crops differently to avoid loss of productivity. Water managers must consider both less groundwater recharge and annual runoff, and asset managers will face increased risks of extreme rainfall events. Biodiversity managers must allow for the complex implications of rapidly changing climate on ecosystems, and increased risks of bushfires and coastal flooding must also be managed (McInnes *et al.* 2003; Suppiah *et al.* 2006). The potential risks create an immediate need to ensure that regional NRM systems are analysed in relation to their vulnerability to climate change impacts, and adaptation needs are responded to by all actors in the governance process (Adger *et al.* 2005; Pelling and High 2005; Füssel and Klein 2006; Campbell 2008).

The research here was undertaken in a partnership between the Federal and South Australian Governments, the Adelaide and Mt Lofty Ranges Natural Resource Management Board and the regional community to examine approaches to generate information to support regional adaptation decision making. The Adelaide and Mt Lofty Ranges Natural Resource Management (AMLR NRM) Region is one of eight NRM regions in South Australia (SA) that were defined under the *Natural Resources Management Act (SA) 2004* (Government of South Australia 2004; AMLR NRM Board 2008a). The climate of the AMLR region and much of SA is Mediterranean. Climate projections from the Intergovernmental Panel on Climate Change (IPCC) and others indicate that Mediterranean climate types are more likely than other climate systems to experience a future drying trend (Fu *et al.* 2006; IPCC 2007). The AMLR region is of special significance for climate change planning in SA because while it remains a terrestrial 'island' of relatively cool, moist environmental conditions in a 'sea' of aridity, it acts as a refuge for biodiversity and a key area of natural resource exploitation for the state (Suppiah *et al.* 2006).

Recognising that climate change adaptation responses can no longer be considered separately from other NRM activities, the stakeholders worked in partnership to develop mechanisms to learn to adapt, following several important steps including:

1. developing an adaptation decision-making framework to support natural resource managers;
2. undertaking an assessment of key areas of regional NRM that are vulnerable to climate change;
3. researching community perceptions of climate change impacts, and
4. developing and demonstrating methods to assist natural resource managers to address climate change risk and develop adaptation responses in vulnerable sectors.

1. Adaptation framework

The sheer enormity of the complex climate change challenges for NRM and the uncertainty of climate projections can be overwhelming enough to lead to paralysis of action. It is important to recognise that in each place there will need to be a process of learning to adapt to climate change and that this will involve numerous steps (Berkhout *et al.* 2006; Kok and de Coninck 2007). The framework developed by this project (outlined below) shows a clear and logical progression of how NRM managers can develop climate change adaptation strategies in their regions. The steps followed are:

1. raising awareness and ownership of climate change;
2. conducting vulnerability assessments of the region;
3. developing adaptation responses, and
4. rigorous analysis of and integration of appropriate adaptation responses into management and planning activities across different time frames, including:
   a. incorporation of climate change into risk management approaches in the short-term;
   b. application of adaptive management techniques that can be adjusted over time, and
   c. application of decisions based on the precautionary principle that allow for increased long-term risk.

5. ongoing revision, reassessment and alteration of those approaches including cost–benefit analysis reviews.

In this case, the precautionary principle definition applied should be as defined by the federal government, which suggests that where there is a lack of full scientific certainty, this should not be used as a reason for postponing a measure to prevent degradation of the environment where there are threats of serious or irreversible environmental damage (Australian Government 2007).

2. Regional vulnerability assessment

The vulnerability of NRM systems in the AMLR was analysed using a methodology outlined by the Allen Consulting Group (2005), which involves an examination of exposure and sensitivity to climate change and associated environmental changes, and an analysis of the capacity for systems to adapt successfully to those changes (Campbell 2008). The vulnerability assessment indicated that there is significant capacity for adaptation to potential climate change by 2030 (Bardsley 2006). In broad terms, the most vulnerable systems were initially assessed to be those that are under less human management and control such as biodiversity conservation, or those that have long management response time frames, namely coastal and bushfire management, biodiversity conservation and perennial horticultural systems. Other systems, particularly water and land management, will require significant human intervention to reduce their vulnerability, particularly within important local catchments and the neighbouring Murray–Darling Basin.

The integrated analysis of vulnerabilities is beginning to assist in the development of planning responses to climate change in the region. For example, the AMLR NRM Board is undertaking projects which respond directly to identified vulnerabilities for biodiversity conservation, coastal management and other areas over the next three years (AMLR NRM Board 2008b). Local councils have also used the work to begin to develop local responses to key concerns such as land use planning, biodiversity conservation and flooding (for relevant examples, see Caton 2007; City of Onkaparinga 2007; City of Port Adelaide Enfield 2007; Adelaide Hills Council 2008).

3. Community perceptions research

A significant process of community awareness-raising and engagement was undertaken initially, and has been described elsewhere (Bardsley and Rogers, forthcoming). Research to understand how key stakeholders perceive climate risk was also seen as important to ensure that methods are employed to engage the NRM community, to identify requirements for skills and knowledge development, and to help engender community ownership of management responses to change (Bardsley and Liddicoat 2008). This work was undertaken at the same time as a series of detailed case studies that formed the basis of the analysis of adaptation response options. Future work to develop effective, governance structures that allow for adjustment of management systems to the changing conditions will only be possible if strong relationships are maintained between different decision-making bodies, and the research and monitoring communities.

4. Develop and assess adaptation responses

It was recognised by AMLR NRM stakeholders that considerably more integrated research was needed in the region to guide specific short- and long-term planning goals, as described in the decision framework. Six detailed case studies targeting climate change impacts relevant to some of the more vulnerable NRM systems were developed across the region (see Figure 15.1).

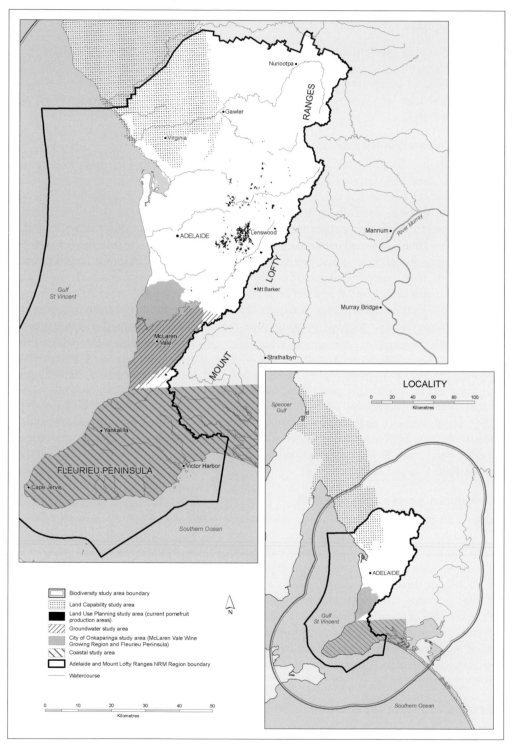

Figure 15.1: Map of the AMLR NRM region showing the case study areas. (From Bardsley and Sweeney 2008)

The case studies were designed to test the application of the decision-making framework via the application of four key research approaches for developing adaptation responses, namely: scenario modelling; applied and participatory GIS modelling; environmental risk analysis, and participatory action learning. The different research approaches to guide decision making represent a spectrum from those that rely strongly on science-led analyses and scenario modelling, through to stakeholder-led participatory research (Dessai *et al.* 2005).

A scenario modelling approach was applied to the Land Capability (DWLBC 2008) and Groundwater studies (Waclawik 2007). These aimed to adjust previous resource condition assessments according to potential scenarios to raise awareness of the types of sectoral impacts that might be expected from climate change. The Land Capability case study aimed to develop and test a scenario modelling approach to determine and analyse the possible impacts of projected climate change on the potential for soil erosion and land degradation. The model was developed for an area primarily north of the AMLR NRM region where there is a contiguous area of land subject to wind erosion (DWLBC 2008). The results identified land where the potential for wind erosion may increase, and located the most vulnerable areas of land susceptible to wind erosion. Appropriate land management practices are suggested in the report to offset any increase in the potential for erosion. Similarly, the discussion paper on the potential impact of climate change on the groundwater resources of the McLaren Vale Prescribed Wells Area (PWA) developed future scenarios and analysed potential impacts of climate change on the groundwater resources of the area (Waclawik 2007). The work used the McLaren Vale PWA groundwater flow model to estimate the impact of varying the rainfall recharge rate and to describe the implications for the groundwater resource. Modelling results confirmed that previous management responses have improved the resilience of the groundwater system, although the impact of climate change on groundwater-dependent ecosystems was identified as a potential significant concern.

Applied and participatory GIS modelling approaches were applied for the Biodiversity (Crossman *et al.* 2008) and Land Use Planning case studies (Houston and Rowland 2008). These aimed to maximise engagement with the stakeholders in the different sectors as modelling was developed, so that key vulnerabilities could be further highlighted and responses articulated. The modelling of native and exotic flora distributions under climate change mapped changes to bioclimatic envelopes for identified species at risk due to climate change (Crossman *et al.* 2008). Stakeholder surveys and feedback workshops, involving interactive discussion were used to analyse the results and influence the revision of further modelling. The work identified which species are most sensitive and where their current habitat is likely to shift in response to climate change. It also examined management opportunities to respond to those changes in relation to environmental planning, ecological restoration prioritisation and invasive species management. The Land Use Planning project also applied a participatory GIS approach to examine key vulnerabilities to climate change, and looked at the apple industry in the AMLR NRM region. An interactive GIS model was developed, in consultation with industry representatives and specialists, to provide a better understanding of land suitability and resource availability for apple production under current and future climate conditions. Opportunities for adaptation in relation to land use planning and management activities continue to be examined with government and industry bodies.

The Perennial Horticulture case study utilised the third approach, a formal environmental risk analysis approach, to guide stakeholders through an analysis of their NRM systems (James and Liddicoat 2008). The formal risk assessment approach applies a matrix with impact and likelihood components included, in contrast to the vulnerability assessment process which does not have a likelihood component. The environmental risk assessment was undertaken to develop climate change adaptation strategies for two key horticultural industry groups in the AMLR region: the McLaren Vale grape growers and Fleurieu Peninsula olive growers. The City of Onkaparinga took the lead role in implementing the case study and engaging with key stakeholders and it is utilising the results to guide planning processes.

Finally, the Coastal study utilised a participatory action learning approach, by which stakeholders are able to analyse their local vulnerabilities in some greater depth without significant information from external sources such as scientific or other public materials (Raymond 2008). The project compared public perception and expert assessment of conservation values and the threats posed by climate change in the Southern Fleurieu Peninsula, with the goal of informing local climate change adaptation responses.

There were advantages and disadvantages to each of the approaches used in the different contexts and there are some major points to consider when examining the appropriate approach to use to develop adaptation responses in any particular context (Table 15.1).

The scenario modelling approach successfully reinforced or improved the knowledge about climate change effects and potential adaptation responses, particularly where resource use is currently marginal. It was also useful to indicate where a broadening of the resilience to change had occurred in existing management systems. Improvements in water-use efficiencies and recycling at McLaren Vale have significantly reduced the vulnerability of supply and hence lowered the area's vulnerability to possibly longer and more frequent dry periods. Such modelling approaches can be criticised, however, in that the focus is on a subset of the range of possible implications of change, thus the modelled outputs are limited by incomplete knowledge of current systems. A further limitation to scenario modelling, also true of the applied and participatory GIS modelling approach, is that systems are so poorly understood that the validity of future scenarios can be undermined.

When there is considerable uncertainty in both future climate and resource conditions, significant stakeholder input can reinforce the legitimacy of strong recommendations. The important involvement of stakeholders in the development of knowledge during the Biodiversity and Land Use Planning studies suggest that participation can also lead to broader outcomes such as greater understanding of the challenges to be faced, and the need for planning to become more explicit about future adaptation needs. The use of participatory modelling processes, which are repeatable, justifiable and have involved critical input from regional stakeholders, supports the development of convincing arguments for better protection of key spaces in the landscape or stricter regulation of resources at risk.

The environmental risk assessment, which framed the Perennial Horticulture case study, takes the participatory approach a step further by engaging stakeholders fully in the development and analysis of risk criteria. However, while the risk assessment process identified immediate and valid concerns for the McLaren Vale grape and olive growers, the framework struggled to guide the required broader examination of NRM issues over the longer term. There is the particular concern that even regions with significant capacity to adapt to climate change will continue to rely upon short-term responses to risk, and will fail to ensure that transformative processes are put in place to accommodate significant change.

The final approach recognises specifically the validity of an argument that to manage the uncertainty of climate projections, a broadening of the resilience and flexibility of management systems is necessary to cope with whatever the future may bring. The participatory action learning approach can be applied to empower stakeholders to analyse and use information, in combination with their own knowledge, to understand the implications of climate change on their own systems much better. As such, it aims to engage stakeholders to imagine how complex long-term risks will affect valued NRM assets. While the approach benefits from

Table 15.1   Suggested use of approaches to guide decision-making and development of climate change adaptation

	Case study			
	Land capability, Groundwater	Biodiversity, Land use planning	Perennial horticulture	Coastal
Adaptation approach				
	Scenario modelling	Applied and participatory GIS modelling	Environmental risk analysis	Participatory action learning
Case study process	Adjust resource condition assessments according to potential climate change scenarios to raise awareness of potential impacts and develop appropriate adaptation responses.	Maximise engagement with industry stakeholders and natural resource managers as modelling is developed, so that key vulnerabilities can be further highlighted and responses discussed.	Conduct a formal risk assessment with impact and likelihood components to guide stakeholders through an analysis of their systems.	Work with stakeholders to identify and analyse local vulnerabilities to climate change in absence of information from external sources.
When approach could be used in future studies	When seeking specific guidance to understand vulnerability of a natural resource. Also when good background data is available concerning the NRM issue, but specific climate change implications are uncertain.	When seeking specific guidance to understand vulnerability of a natural resource. Also when the development of good background data concerning the NRM issue requires stakeholder input.	When trying to formally involve stakeholders in a process of analysing risk. Ideally supported with empirical data to inform planning outcomes. Likelihoods and consequences are well understood.	When community support needs to be generated and/or articulated to support difficult decision making. Particularly when seeking to generate greater awareness of climate change risks.

imagining future scenarios, the research for Fleurieu Peninsula was criticised for the lack of empirical rigour, which could make the immediate use of the findings more difficult.

Conclusion

Given that the uncertainties of future natural resource conditions are so considerable, research into local climate change impacts and adaptation options will require considerable flexibility, so that adaptation responses can accommodate the potential for great change to landscapes, systems and society. Consistent with good NRM practice anywhere, the industry and community representatives involved in this project identified that the sustainable management of future landscapes in light of climate change will require a mix of research and education, coupled with planning policies that incorporate both incentives and regulation, and are flexible over time. Positive, action-based leadership that integrates all levels of NRM and societal governance is necessary to bring about the changes required. Beyond that conclusion, there are specific messages emerging from the project, namely:

- the framework supported decision-making processes and provided a clear sequence of steps for natural resource managers to engage the community and develop climate change adaptation responses;
- the vulnerability assessment process aids in the prioritisation of adaptation planning;
- involvement of stakeholders and key decision makers is critical in developing adaptation responses, particularly where detailed local information on resource or climatic conditions is unavailable or of limited quality;
- information on likely impacts of climate change must include adequate reference to the level of uncertainty, and
- while cost–benefit analyses were not undertaken in this work, more detailed future studies could incorporate the benefits of early action, or costs from a lack of or inappropriateness of responses, to assess the suitability of investment.

Effective regional NRM response strategies must overcome considerable inertia in order to respond to the challenges posed by climate change. That transition may well involve a change in understanding of our NRM regions as places whose characteristics are known, stable and abundant, to places whose characteristics are not fully understood, and which may change significantly over time. The premise of this chapter has been that to guide transformative action in relation to climate change, research needs to examine local understandings of systems, as well as to develop critical knowledge of the information, rules, institutions and systems of management that prevail in the region. Initial studies such as those discussed here provide better sectoral and spatial understandings of risk, but they provide only a snapshot of current information and understanding. Strategic responses will need to be developed and improved upon over time, to ensure that approaches to adaptation reflect the changing nature of risk that regional NRM is facing. Primary to such an adaptive approach is the ownership of the issue by stakeholders, which leads to an acceptance that actions to adapt and mitigate are now both possible and vitally important.

Acknowledgements

This work was supported financially by the Department of Water, Land and Biodiversity Conservation, the Australian Government through the Department of Climate Change, (formerly the Australian Greenhouse Office in the Department of the Environment and Water Resources), and by the AMLR NRM Board. Many thanks to the AMLR NRM community for their ongoing support of this project.

References

Adelaide Hills Council (2008) 'Adelaide Hills Council Development Plan: Flood management'. Draft Development Plan Amendment by Adelaide Hills Council, January 2008. <www.ahc.sa.gov.au/webdata/resources/files/Draft_AHC_Flood_Management_DPA_Version_7.pdf>

Adger WN, Arnell NW and Tompkins E (2005) Successful adaptation to climate change across scales. *Global Environmental Change* **15**, 77–86.

Allen Consulting Group (2005) 'Climate change risk and vulnerability: Promoting an efficient adaptation response in Australia'. Commonwealth of Australia, Canberra.

AMLR NRM (Adelaide Mt Lofty Ranges Natural Resource Management) Board (2008a) 'Creating a sustainable future: an integrated natural resources management plan for the Adelaide and Mount Lofty Ranges Region'. Volume A – State of the region report. South Australian Government, Adelaide.

AMLR NRM (Adelaide and Mt Lofty Ranges Natural Resource Management) Board (2008b) 'Creating a sustainable future: a natural resources management plan for the Adelaide and Mount Lofty Ranges Region'. Volume C – the board's investment plan 2008/09–2010/11. South Australian Government, Adelaide.

Australian Government (2007) *Environment Protection and Biodiversity Conservation Act 1999: Act Compilation (superseded) - C2007C00385*. Commonwealth of Australia, Canberra.

Bardsley D (2006) 'There's a change on the way: an initial integrated assessment of projected climate change impacts and adaptation options for natural resource management in the Adelaide and Mt Lofty Ranges region'. Department of Water, Land and Biodiversity Conservation Report 2006/06. South Australian Government, Adelaide.

Bardsley DK and Liddicoat C (2008) 'Community perceptions of climate change impacts on natural resource management in the Adelaide and Mount Lofty Ranges'. Department of Water, Land and Biodiversity Conservation Report 2008/14. South Australian Government, Adelaide.

Bardsley DK and Rogers G (In press) Prioritising engagement for sustainable adaptation to climate change: an example from natural resource management in South Australia. *Society and Natural Resources*.

Bardsley DK and Sweeney S (2008) 'A regional climate change decision framework for natural resource management'. Department of Water, Land and Biodiversity Conservation Report 2008/21. South Australian Government, Adelaide.

Berkhout F, Hertin J and Gann DM (2006) Learning to adapt: organisational adaptation to climate change impacts. *Climatic Change* **78**, 135–156.

Botkin DB, Saxe H, Araújo MB, Betts R, Bradshaw RHW, Cedhagen T, Chesson P, Davis MB, Dawson TP, Etterson J, Faith DP, Ferrier S, Guisan A, Skjoldborg HA, Hilbert DW, Loehle C, Margules C, New M, Sobel MJ and Stockwell DRB (2007) Forecasting effects of global warming on biodiversity. *Bioscience* **57**, 227–236.

Campbell A (2008) 'Managing Australian landscapes in a changing climate: a climate change primer for regional natural resource management bodies'. Report to the Department of Climate Change, Canberra, Australia.

Caton B (2007) 'The impact of climate change on the coastal lands of the City of Onkaparinga'. Report for the City of Onkaparinga 18 July 2007. <www.onkaparingacity.com/web/binaries?img=15099&stypen=html>

City of Onkaparinga (2007) 'Addressing climate change. Strategic directions 2020 discussion paper 2, June 2007'. <www.onkaparingacity.com/web/binaries?img=9442&stypen=html>

City of Port Adelaide Enfield (2007) 'State of environment report June 2007'. City of Port Adelaide Enfield. <www.portenf.sa.gov.au/site/page.cfm?u=1160>

Crossman ND, Bryan BA and Bardsley DK (2008) 'Modelling native and exotic flora distributions under climate change'. CSIRO Land and Water Science Report 01/08. CSIRO, Adelaide.

Dessai S, Lu X and Risbey JS (2005) On the role of climate scenarios for adaptation planning. *Global Environmental Change* **15**, 87–97.

DWLBC (Department of Water, Land and Biodiversity Conservation) (2008) 'Climate Change and the potential for wind erosion – a model for the Adelaide and Mt Lofty Ranges NRM region'. South Australian Government, Adelaide.

Fu Q, Johanson CM, Wallace JM and Reichler T (2006) Enhanced mid-latitude tropospheric warming in satellite measurements. *Science* **312**, 1179.

Füssel H-M and Klein RJT (2006) Climate change vulnerability assessments: An evolution of conceptual thinking. *Climatic Change* **75**, 301–329.

Government of South Australia (2004) *South Australia Natural Resources Management Act 2004*. South Australian Government, Adelaide.

Houston P and Rowland J (2008) 'Room to move: towards a strategy to assist the Adelaide Hills apple industry adapt to climate change in a contested peri-urban environment'. Department of Water, Land and Biodiversity Conservation Report 2008/20. South Australian Government, Adelaide.

Hughes L (2003) Climate change and Australia: trends, projections and impacts. *Austral Ecology* **28**, 423–443.

IPCC (2007) Climate Change 2007: The Physical Science Basis. Contribution of Working Group I to the Fourth Assessment Report of the Intergovernmental Panel on Climate Change. Cambridge University Press, Cambridge, UK.

James J and Liddicoat C (2008) 'Developing industry climate change adaptation strategies: a case study for the McLaren Vale viticulture and Fleurieu Peninsula oliveculture industries'. Department of Water, Land and Biodiversity Conservation Report 2008/11. South Australian Government, Adelaide.

Kok MTJ and de Coninck HC (2007) Widening the scope of policies to address climate change: Directions for mainstreaming. *Environmental Science & Policy* **10**, 587–599.

McInnes KL, Suppiah R, Whetton PH, Hennessy KJ and Jones RN (2003) 'Climate change in South Australia'. CSIRO Atmospheric Research, Melbourne.

Pelling M and High C (2005) Understanding adaptation: what can social capital offer assessments of adaptive capacity? *Global Environmental Change* **15**, 308–319.

Potter NJ, Chiew FHS, Frost AJ, Srikanthan R, McMahon TA, Peel MC and Austin JM (2008) 'Characterisation of recent rainfall and runoff in the Murray-Darling Basin'. A report to the Australian Government from the CSIRO Murray-Darling Basin Sustainable Yields Project. CSIRO, Australia.

Raymond C (2008) 'Mapping landscape values and perceived climate change risks for natural resources management: a study of the Southern Fleurieu Peninsula region, SA'. Department of Water, Land and Biodiversity Conservation Report 2008/07, South Australian Government, Adelaide.

Suppiah R, Preston B, Whetton PH, McInnes KL, Jones RN, Macadam I, Bathols J and Kirono D (2006) 'Climate change under enhanced greenhouse conditions in South Australia'. CSIRO, Australia.

Waclawik V (2007) 'Discussion Paper on the Potential Impact of Climate Change on the Groundwater Resources of the McLaren Vale Prescribed Wells Area'. Department of Water, Land and Biodiversity Conservation, South Australian Government, Adelaide.

16. RESPONDING TO OIL VULNERABILITY AND CLIMATE CHANGE IN OUR CITIES

Peter Newman

Abstract

Reducing oil use is a somewhat neglected part of climate change policy. This paper examines how opportunities exist for substantial reductions through electric transit and electric plug-in vehicles all powered by renewable energy, through natural gas and biofuels in freight and regional transport, through telepresence, high speed rail and airships, all of which enable the economy to be less oil-dependent and hence less greenhouse gas-intensive in its transport system.

The problem

This paper will focus on how we can reduce the oil-derived component of greenhouse gases. Oil contributes around one-third of the global problem arising from fossil fuel use and seems to be the area most out of control; i.e. until recently transport was growing faster than any other area of fuel use. The majority of oil is used in cities and hence we have to rethink our cities. Much climate change policy in cities is about adapting to rising sea levels, but the causes of oil dependence must be addressed or else there is little point in trying to fix vulnerable coastal parts of cities. Solving this problem would involve multiple benefits beyond the impact on climate change, including reducing peak oil vulnerability, smog emissions, and improving human health, safety and equity.

The opportunity

The response to the challenge of reducing any fossil fuel use dramatically can often be one of panic – the fear that it will have a severe impact on our economy. It is also possible to see that this is a real opportunity and that indeed the ability of cities to compete effectively in reducing oil use will be a major part of their new economies (Newman *et al.* 2009). The next phase of innovation in our cities is seen by some to be based around sustainability innovations as set out by Hargraves and Smith (2004) in Figure 16.1. It is worth noting that these innovation cycles usually grow out of a period of economic downturn (e.g. the coal-to-oil transition occurred in the 1930s), and we can expect significant economic disruption when the next transition occurs from oil to renewables and smart cities.

Radical resource productivity and renewable energy can be seen to be linked to the digital networks of the previous innovation phase to produce a whole new set of economic

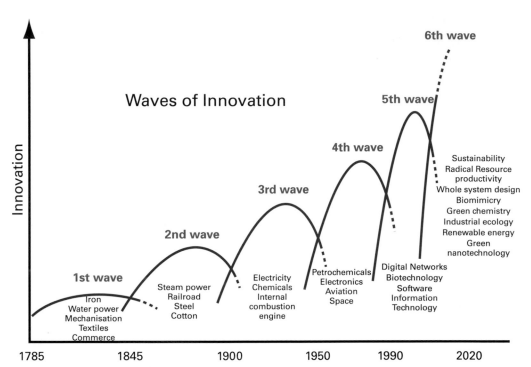

Figure 16.1: Waves of innovation experienced over recent history. The sixth wave of sustainablity, complex systems and renewable and green technologies is developing rapidly.

opportunities. These opportunities will make cities and their regions overcome their oil addictions and move towards a much greater degree of resilience. The question from this perspective then becomes: what are these opportunities and how can we respond best to them? This chapter will present four areas in which transport opportunities could help to address the problem of oil-dependency:

1. new generation electric transit systems and their associated transit-oriented development (TOD), pedestrian-oriented development (POD) and green-oriented development (GOD) structures;
2. renewable energy-based electric vehicles linked through 'smart grids';
3. natural gas and biofuels in freight and regional transport, and
4. telepresence, high speed rail and airships.

1. New generation electric transit systems and their associated TOD, POD and GOD structures

Cities need to have a combination of transportation and land use options that are favorable for green modes and offer a time saving when compared to car travel. This means transit needs to be faster than traffic down each major corridor. Those cities where transit is relatively fast are those with a reasonable level of support for it. The reason is simple – they can save time.

With fast rail systems, the best European and Asian cities have achieved a transit option that is faster than the car down the main city corridor. Rail systems are faster in every city in our 84 city sample by 10 to 20 kilometres per hour (kph) over bus systems, as buses rarely average over 20 to 25 kph (Kenworthy and Laube 2001; Kenworthy *et al.* 1999). Busways with a designated lane can be quicker than traffic in car-saturated cities, but in lower density car-dependent cities it is important to use the extra speed of rail to establish an advantage over cars in traffic. This is one of the key reasons why railways are being built in over 100 US cities.

Rail has a density-inducing effect around stations, which can help to provide the focused centres so critical to overcoming car dependence. Thus transformative change of the kind that is needed to rebuild car-dependent cities comes from new electric rail systems, as they provide a faster option than cars and can help build transit-oriented centres.

It goes without saying that we need to stop increasing road capacity if we are going to begin this transition to a green city and its associated economy. This is not easy given the funding systems that have developed around the car and truck. However evidence from the US shows that car use is beginning to spiral downwards and last year had its biggest drop for 50 years at 4.3% per year. Meanwhile transit use increased 6.5% – also a record (Puentes and Tomer 2009).

How much can we change our cities? It is possible to imagine an exponential decline in car use in our cities that could lead to 50% less driving (passenger kilometres). The key mechanism is a quantitative leap in the quality of public transport whilst fuel prices continue to climb, accompanied by an associated change in land use patterns.

Figure 16.2 shows the relationship between car passenger kilometres and public transport passenger kilometres from the CUSP Global Cities Database. The most important thing about this relationship is that as the use of public transport increases linearly the car passenger kilometres decrease exponentially. This is due to a phenomenon called transit leverage whereby one passenger kilometre of transit use replaces between three and seven passenger kilometres in a car due to more direct travel (especially in trains), trip chaining (doing various other things like shopping or service visits associated with a commute), giving up one car in a

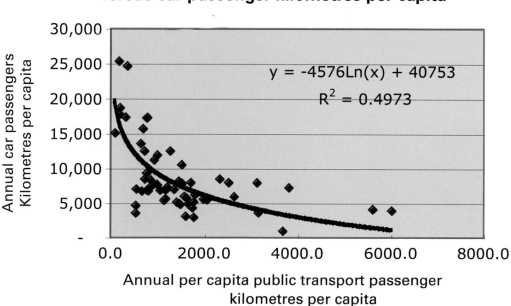

Public transport passenger kilometres per capita *versus* car passenger kilometres per capita

$$y = -4576Ln(x) + 40753$$
$$R^2 = 0.4973$$

(y-axis) Annual car passengers Kilometres per capita

(x-axis) Annual per capita public transport passenger kilometres per capita

Figure 16.2: Comparison of the annual per capita public transport kilometres travelled compared to the kilometres travelled per capita in a motor vehicle.

household (a common occurrence that reduces many solo trips) and eventually changes in where people live as they prefer to live or work nearer transit (Newman and Kenworthy 1999).

If Sydney doubled its transit use to 3018 passenger kilometres per person it would (from Figure 16.2) have a per capita private transport use of 4088 passenger kilometres per capita which is a 61% reduction in car passenger kilometres per person over the 1996 figure. If Perth was able to continue the rapid growth in transit patronage and triple its 1996 use to around 2000 passenger kilometres per person then it would reduce its private transport use per capita to 6000 car passenger kilometres per capita, which is a reduction of 56% over the 1996 level. Similar calculations can be done for the other Australian cities. These are not difficult to imagine as they represent growth rates of around 2% per year for transit.

The biggest challenge in an age of radical resource efficiency requirements will be finding a way to build fast rail systems for the scattered car-dependent cities of Australia. How can a fast transit service be built back into these areas? The solution may well be provided by Perth and Portland which have both built fast rail systems down freeways. Freeways are public facilities that may well be in decline in the future as car traffic faces the double whammy of increasing fuel prices due to peak oil and carbon taxes due to climate change. To build fast electric rail down the middle of these roads is easier than anywhere else as the right of way is there and engineering in terms of gradients and bridges is compatible. They are not ideal in terms of ability to build transit-oriented developments (TOD), but building them can still be done using high-rise buildings as sound walls. Links from buses, electric bikes, and park-and-ride are all easily provided so that local travel to the system is short and convenient. The key is the speed of the transit system, and in Perth the new Southern Railway has a maximum speed of 130 kph (80 mph) and an average speed of 90 kph (55 mph) which is at least 30% faster than traffic. The result is dramatic increases in patronage (55 000 passengers per day compared to 14 000 on the buses) far beyond the expectations of planners who see such suburbs as too low in density to deserve a rail system. There is little else that can compete with this kind of option for creating a future in the car dependent suburbs of many cities.

Fast electric rail services are not cheap. However, they cost about the same per kilometre as most freeways and we have been able to find massive funding sources for these in the past 50 years. In the transition period it will require some creativity as the systems for funding rail are not as straightforward, which is one of the reasons for establishing Infrastructure Australia (the new funding body designed to improve the productivity and sustainability of Australian cities and regions). In the 2009/10 budget announcement, there was a $4.6 billion boost for urban passenger rail, an historic first for a federal government budget.

Transit-oriented development (TOD) has become a major technique for reducing automobile dependence and hence tackling peak oil. The facilitation of TODs has been recognised by all Australian and many American cities in their metropolitan strategies, which have developed policies to reduce car dependence through centres along corridors of quality transit. This can be a significant source of funding for the required rail infrastructure through 'Value Transfer PPPs' (public–private partnerships), as in the very successful Chatswood Transport Interchange PPP which has created a new railway station and bus interchange along with a retail and residential complex that makes a small city around and over the station (Dawson 2008).

The major need for TODs is not in the inner areas, as these have many from previous eras of transit building; however, the newer outlying suburbs, built in the past four or five decades, are heavily car-dependent with high fuel consumption and almost no TOD options available. There are real equity issues here as the poor are increasingly trapped on the fringe with high expenditures on transport. A 2008 study by the Center for Transit Oriented Development shows that people in TODs drive 50% less than those in conventional suburbs and save 20% of their household income as a result. In both Australia and the USA, homes that are located in TODs are holding their value the best or appreciating the fastest under the pressure of rising fuel prices (Cervero 2008).

TODs must also be PODs (pedestrian-oriented development) or they lose their key quality as a car-free environment to which businesses and households are attracted. This is not automatic but requires the close attention of urban designers. Jan Gehl's transformations of central areas such as Copenhagen and Melbourne are showing the principles of how to improve TOD spaces so they are more walkable, economically viable, socially attractive, and environmentally significant (Gehl Architects 2009). It will be important for those green developers wanting to claim credibility that scattered urban developments, no matter how green in their buildings and renewable infrastructure, will be seen as failures in a post-peak oil world unless they are building pedestrian-friendly TODs.

At the same time TODs that have been well designed as PODs will also need to be GODs (green-oriented developments). TODs will need to ensure that they have full solar orientation, are renewably powered with smart grids (digitally-controlled electricity grids to enable efficiency and precision with fluctuating renewable inputs), have water-sensitive design, use recycled and low-impact materials, and incorporate innovations like green roofs.

Perhaps the best example of a TOD-POD-GOD is the redevelopment of Kogarah Town Square in Sydney. This inner city development is built upon a large City Council carpark adjacent to the main train station where there was a collection of poorly performing businesses adjacent. The site is now a thriving mixed-use development consisting of 194 residences, 50 000 square feet of office and retail space and 35 000 square feet of community space including a public library and town square. The buildings are oriented for maximum use of the sun with solar shelves on each window (enabling shade in summer and deeper penetration of light into each room), photovoltaic (PV) collectors are on the roofs, all rainwater is collected in an underground tank to be reused in toilet flushing and irrigation of the gardens, recycled and low-impact materials were used in construction, and all residents, workers, and visitors to the site have a short walk to the train station (hence reduced parking requirements enabled better and more productive use of the site). Compared to a conventional development, the Kogarah Town Square saves 42% of the water and 385 tonnes of greenhouse gas. This does not include transport oil savings that are hard to estimate but are likely to be even more substantial.

2. Renewable energy-based electric vehicles linked through smart grids

Even if we manage to reduce car use by 50%, we still have to reduce the oil and carbon in the other 50% of vehicles being used. The question should therefore be asked: what is the next best transport technology for motor vehicles? The growing consensus seems to be plug-in electric hybrid vehicles (PHEV). Plug-in electric vehicles are now viable alternatives due to the new batteries such as lithium ion, and with hybrid engines for extra flexibility, they are likely to be attractive to the market. The key issue here is that plug-in electric vehicles not only reduce oil vulnerability but may become a critical component in how renewable energy is incorporated into a city's electricity grid. The PHEVs will do this by enabling renewables to have a storage function.

After electric vehicles are recharged at night, they can be a part of the peak power provision next day when they are not being used but are plugged in. Peak power is the expensive part of an electricity system and suddenly renewables are offering the best and most reliable option. Hence the resilient city of the future is likely to have a significant integration between renewables and electric vehicles through a smart grid. Thus electric buses, electric scooters and gophers, and electric cars have an important role in the future resilient city, both in helping to make its buildings renewably powered and in removing the need for oil in transport (Went *et al.* 2008).

Electric rail can also be powered from the sun either through the grid powering the overhead wires or in the form of new light rail (with these new Li-Ion batteries) which could be built down highways into new suburbs without requiring overhead wires. Signs that this transition to electric transport is underway are appearing in USA-based demonstration projects such as those in Boulder and Austin, in Google's 1.6 megawatt (MW) solar campus in California (with 100 PHEVs) and by the fact that oil companies are acquiring electric utilities. The Obama stimulus package contains support for this technology.

What sort of impact could these kinds of integrated systems have on the electricity supply system? According to one study the integration of hybrid cars with the electric power grid could reduce gasoline consumption by 85 billion gallons per year which is equal to:

- 27% reduction in total US greenhouse gases;
- 52% reduction in oil imports, and
- $270 billion not spent on gasoline (Kinter-Meyer *et al.* 2007).

The real test of a resilient city will be how it can simultaneously be reducing its global greenhouse and oil impact through these new technologies whilst reducing the need to travel by car through the policies outlined in the first strategy on transit and TODs.

3. Natural gas and biofuels in freight and regional transport

What do you do with freight transport and regional transport outside cities where electric grids are not so easily used with vehicles?

There will almost certainly be a reduction in the amount of freight moving around as fuel prices eat into the transport economics of consumption. Container transport will be reduced as fuel costs move from being 10–15% to over 50% of total costs. Food miles will start to mean something to food prices when the cost of fuel triples. But trucks and trains and regional transport will still exist.

The next stage for larger vehicles and for regional transport would appear to be to switch to greater use of natural gas and biofuels. Trucks and trains and fishing boats can use CNG (compressed natural gas) or LNG (liquefied natural gas) in their diesel engines (with pay-off times of just a few years due to high diesel costs). Cars can be switched over as well (particularly if the manufacturer makes them standard as in Sweden, when the Government committed to natural gas cars for their vehicle fleet). The attraction is that natural gas is already in place in terms of infrastructure, although actual filling stations are not commonplace. The benefit of the transition to natural gas is that it enables the long-term transition to hydrogen or renewable natural gas based on solar photocatalysis of CO_2 and water.

Biofuels began by promising a lot but their image has become rather tarnished since delivery began. This is due to their impact on food prices as a result of converting grain into fuel, and to estimates suggesting biofuels may be worse than oil when it comes to climate change. They still have a potentially significant role in some areas (e.g. where there is surplus sugar), and technology may eventually make it possible to create biofuels from cellulose materials (agricultural and forestry waste) and blue-green algae. It is likely that biofuels will be used as a do-it-yourself fuel on farms. Thus biofuels may have a role in agricultural regions as a fuel to assist farmers in their production but they are not yet options that can be taken seriously as a widespread fuel for cities.

4. Telepresence, high speed rail and airships

Transport to meet people by long distance or even short distance trips within cities may not be needed once the use of broadband-based telepresence begins to make high quality imaging feasible on a large scale. There will always be a need to meet face-to-face in creative meetings in cities, but for many routine meetings the role of computer-based meetings will rapidly take off (Newman *et al.* 2009).

Aircraft are not going to cope easily with the rapid rise in fuel prices. At the height of the 2008 fuel crisis Garry Kelly, the CEO of Southwest Airlines said, 'No airline can make money at $123 a barrel' (Seaney, 2009). Gilbert and Perl (2007) suggest a few ways that air travel will adapt but they see little potential other than regional high speed rail and a return to ship travel.

Perhaps the technology that could make a comeback is airships. These are able to fly at low levels at speeds of 150–200 kph and carry large loads with one-tenth the fuel of aircraft technology. They are already being used to carry large mining loads to remote areas and to take groups of around 200 on ecotourism ventures similar to a cruise ship.

Conclusions

There are not many guidelines for the future of our cities and regions that take account of what could happen to transport in response to climate change and peak oil. It is understandable therefore why some people get very upset about the possibilities of collapse. As Lankshear and Cameron (2005) say: 'Peak oil has already become a magnet for post-apocalyptic survivalists who are convinced that western society is on the brink of collapse, and have stocked up tinned food and ammunition for that coming day.'

The alternatives all require substantial commitment to change in both how we live and the technologies we use in our cities and regions. The need to begin the changes is urgent as they will take decades to get in place, and the time to respond to peak oil and climate change is of the same order. But at least by imagining some of the changes as suggested above it is possible to see how we can get started on the road to more resilience and sustainability in our settlement transport systems.

References

Cervero R (2008) Transit-oriented development in America: strategies, issues, policy directions. In: *New Urbanism and Beyond: Designing Cities for the Future.* (Ed. T Haas) pp. 124–129. New York, Rizzoli.

Dawson B (2008) The new world of Value Transfer PPPs. *Infrastructure: Policy, Finance and Investment,* **May** 12–13.

Gehl Architects (2009) *Perth 2009: Public Spaces and Public Life.* Gehl Architects ApS.

Gilbert R and Perl A (2008) *Transport Revolutions: Making the Movement of People and Freight Work for the 21st Century.* Earthscan, London.

Hargraves C and Smith M (2004) *The Natural Advantage of Nations.* Earthscan, London.

Kenworthy J and Laube F (2001) *The Millennium Cities Database for Sustainable Transport.* UITP, Brussels.

Kenworthy J, Laube F, Newman P, Barter P, Raad T, Poboon C and Guia B (1999) *An International Sourcebook of Automobile Dependence in Cities, 1960–1990.* University Press of Colorado, Boulder.

Kintner-Meyer M, Schneider K and Pratt R (2007) *Impacts Assessment of Plug-in Hybrid Vehicles on Electric Utilities and Regional U.S. Power Grids, Pt1: Technical Analysis.* Pacific Northwest National Laboratory, US DoE, DE-AC05-76RL01830.

Lankshear D and Cameron N (2005) Peak oil: a Christian response. *Zadok Perspectives* **88**, 9–11.

Newman P, Beatley T and Boyer H (2009) *Resilient Cities: Responding to Peak Oil and Climate Change.* Island Press, Washington DC.

Newman P and Kenworthy J (1999) *Sustainability and Cities: Overcoming Automobile Dependence.* Island Press, Washington DC.

Puentes R and Tomer A (2009) 'The Road Less Travelled: An Analysis of Vehicle Miles Traveled'. Trends in the U.S. Metropolitan Infrastructure Initiative Series. Brookings Institution, Washington DC.

Seaney R (2009) *High Oil Prices Spell Disaster for Airlines*, ABC News Internet Ventures. <http://a. abcnews.com/m/screen?id=4847008&pid=74>

Went A, James W and Newman P (2008) 'Renewable Transport'. *CUSP Discussion Paper* 1/2008. <www. sustainability.curtin.edu.au/publications>

17. MANAGING CLIMATE RISK IN HUMAN SETTLEMENTS

Benjamin Preston and Robert Kay

Abstract

Risk management has seen widespread use in Australia in recent years as a framework for assessing the potential consequences of climate change to human settlements and identifying appropriate management actions. It represents an over-arching framework for assessment and decision making; however, it is not a prescribed methodology. As a consequence, its application in Australia is characterised by significant diversity with respect to problem framing and analytical approaches. Such diversity is a predictable consequence of the highly contextual nature of climate risk, but it also impedes comparisons across locales. It is therefore argued that supporting continued growth in the use of risk management will require some degree of standardisation in a manner that doesn't sacrifice context. A range of additional barriers will also need to be overcome to ensure climate change risk management activities are oriented appropriately, institutions have sufficient capacity to manage risk and the successes and failures of adaptation actions are tracked and reported transparently.

Introduction

Over the course of the 21st century, cities will become the dominant settlement pattern for the Earth's human population (UN 2006). Whereas only 13% of the global population resided in urban areas in 1900, by 2005 that proportion reached 49% and is projected to reach 60% by 2030. In addition, many of the world's human settlements are, in fact, coastal settlements. In the United States, 51% of the 2000 population could be found within 80 km of the coast, a strip of land representing just 13% of the nation's total area (Rappaport and Sachs 2003). In Australia, 85% of the 2001 population resided within 50 km of the coast (ABS 2004), and the majority of the Australia's population growth is concentrated in urban centres and non-urban coastal regions (ABS 2006). Similar statistics are emerging from the developing world as well (Nicholls 1995; Henderson 2002), particularly in Asia where coastal megacities are growing rapidly in response to economic development and internal migration from rural communities to urban centres. Such rapid urbanisation and coastal development forms the socio-economic context in which climate change and its consequences will manifest.

The challenge for managing climate risk to human settlements is predominately three-fold. First, increasing urbanisation, particularly in coastal areas, will expose a growing number of properties and individuals to climatic hazards including extreme heat events, bushfires and coastal storms. For example, Changnon *et al.* (2000) note that the majority of increases in United States' disaster losses over the 20th century were due to socio-economic trends that placed more people and assets in harm's way, such as flood, bushfire and tropical cyclone-prone regions. Second, climate change is projected to exacerbate a range of climatic hazards including increases in the frequency of events and/or the manifestation of events of histori-cally unprecedented severity. Third, decision making around equitable, effective and efficient adaptation policies and measures to cope with climate change is difficult due to the long time scales and significant uncertainties associated with future states (Mahlman 1998). Such diffi-culties are exacerbated by institutional arrangements and policy structures that are ill-equipped to manage complex and dynamic systems (Hajer 2003).

Risk management has been advocated as an approach to climate change adaptation decision making that can assist in overcoming some of these challenges (Jones 2001). To-date, a range of guidance for the application of the risk management paradigm in the context of climate adaptation exists (Willows and Connell 2003; AGO 2006). Such advocacy and the general familiarity of the concept of risk management across a range of public and private institutions have contributed to climate adaptation increasingly framed in the context of risk. This is evi-denced in Australia by the recent explosion of climate change risk assessments that have emerged across a range of geopolitical scales (local to national). While such activity suggests that institutions are orienting themselves rapidly around the challenges of climate adaptation (in part due to the requirement of various Commonwealth Government grant programs to use risk assessment), this also suggests that a critical examination of the role of risk assessment and the application of the risk management paradigm in adaptation would be timely. This chapter summarises some of the recent experience in climate change risk management for human set-tlements and highlights some of the progress in research and assessment tools that have emerged. At the same time, this chapter identifies some of the challenges associated with risk-based approaches to adaptation and, in particular, the many and varied applications of risk assessment and management in support of adaptation.

Assessing climate risk in human settlements

Over the past five years, there has been a substantial growth in the number of local-scale climate change risk assessments, with much of this effort focused on human settlements, with a particular emphasis on climate change risk in Australian coastal population centres. To a large degree, this surge in interest is indicated by the various programs and initiatives which have been established at all levels of government (Figure 17.1), including:

- the Integrated Assessment of Human Settlements (IAHS) initiative – a series of five projects funded through the Australian Department of Climate Change (DCC), which all included significant biophysical impact assessments as well as trialling a range of methodological approaches to assessment. All five projects were completed by early 2009;[1]
- National Coastal Vulnerability Assessment (NCVA) Case Studies – a series of six pilot studies funded by the DCC to illustrate a range of policy issues, risks and threshold sensitivities related to climate change impacts for selected coastal locations.[2] The case studies are part of the so-called 'first pass' national assessment.[3] All NCVA projects are planned for completion during mid-2009;
- Local Adaptation Pathways Program (LAPP) Phases I & II – 32 local government assess-ment projects throughout Australia (including multiple coastal local governments) were selected on a competitive basis for the LAPP Phase I. These projects applied the

Commonwealth's risk management framework (AGO 2006) using an approved suite of commercial providers in an attempt to standardise and assure the quality of project outcomes.[4] LAPP Phase I projects were due for completion by 30 June 2009. A LAPP Phase II has already started, again on a competitive basis, but with an increased emphasis on regional-scale projects. Seven projects have been funded and work is scheduled to start in late 2009, and

- various projects and studies on climate change impacts and adaptation assessment, either directly funded by the Commonwealth or by local government or individual land developers (in some cases as part of development approval processes). For example, in Western Australia, the Vulnerability of the Cottesloe Foreshore to the Potential Impacts of Climate Change project was funded by Emergency Management Australia (CZM Pty and Damara Ltd 2008; Blackmore and Goodwin 2009). Additional studies have been completed for local governments (NATCLIM, 2007), individual water utilities (Howe *et al.* 2005; Kirono *et al.* 2007) and sensitive ecosystems (Pickering 2004).

The aforementioned programs indicate that a plethora of climate change risk assessments will became available in 2009 and 2010. These studies represent a rich body of knowledge that are likely to help inform local governments about climate risk at the local level and will presumably facilitate learning about the demand and scope of adaptation efforts. At the same time, however, these studies also are likely to highlight the constraints and opportunities embedded in the current Australian focus on using risk-based approaches to climate change impact and adaptation assessments on settlements among the range of methods and tools used internationally (Abuodha and Woodroffe 2006; CZM Pty 2008; UNFCCC 2008). For example, some emerging issues from the operational experience in assessing risk at the local scale using the risk-based Commonwealth framework include (CZM Pty and Damara Ltd 2008; Preston *et al.* 2009a) the following.

- Methodological challenges related to:
  - the estimation of the likelihood of adverse consequences in risk assessment studies;
  - identifying critical thresholds for climate change in specified systems and, more generally, quantifying relationships between climate change drivers, and system processes and responses;
  - identifying thresholds for action – setting management thresholds for implementation of adaptation actions;
  - overcoming data gaps and inconsistences, and
  - the general lack of comparability across different assessments.
- Communication challenges related to:
  - aligning stakeholder expectations with methodological and resource limitations;
  - aligning stakeholder terminology to describe risk and vulnerability, as well as methodological approaches to risk assessment, with those of researchers and technical experts;
  - clearly communicating the trade-offs between strategic, but generic adaptation versus specific adaptation options for site-specific problems and issues, and
  - linking risk to adaptation – translating information about risk at the local scale into actions that build capacity and reduce vulnerability.

The emerging body of climate change risk assessments ultimately reflects different, context-specific approaches to coping with the above challenges. Nevertheless, this diversity is also likely to impede attempts to make comparisons across different locales with respect to risk

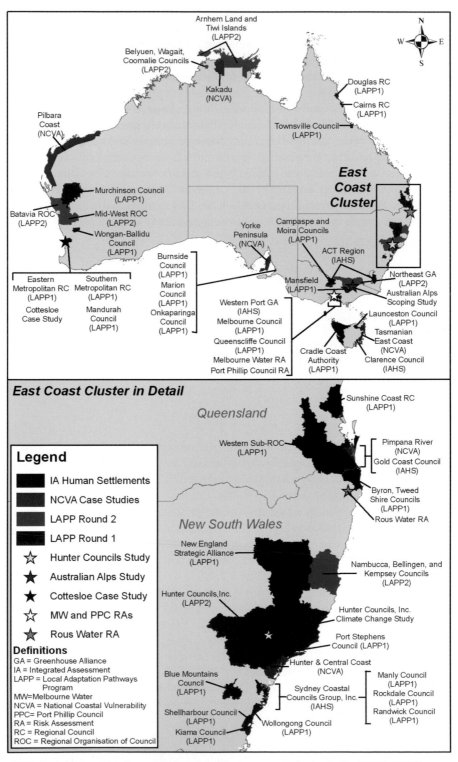

Figure 17.1: Overview of local- and regional-scale assessments in Australia (2004–2009).

due to differences in problem framing, analytical approaches and normative judgements. As such, while risk management may be a broad paradigm for managing decision making around climate adaptation (Jones and Preston 2009), it is important to recognise that a diverse toolbox of methods and approaches may be consistently applied within this paradigm. This includes both top–down and bottom–up approaches, qualitative and quantitative approaches, and predictive and diagnostic approaches. Hence, advocating the use of a risk management approach to adaptation planning (see, for example, AGO 2006) doesn't necessarily ensure a standard methodology will be applied consistently from one location to another. From a policy perspective, however, such standardisation may be desirable, particularly when it comes to making what are likely to be difficult decisions regarding the prioritisation of locations, sectors or issues for interventions. Therefore, it is timely to re-evaluate the relative merits of formalising a more standard methodology for assessment that can be applied across sectors and geopolitical scales versus continuing with the current *ad hoc* approach to assessment.

From risk assessment to adaptation outcomes

Clearly, progress is being made on both assessment and policy development to address climate change risk to Australia's human settlements, albeit in a relatively *ad hoc* fashion that gives the impression of standardisation because it uses the same risk assessment framework, but produces very diverse outcomes. The case studies and pilot projects either being undertaken currently or recently completed at a local scale, have provided important lessons and have clearly stimulated debate. It is likely that there will be increased policy activity in the future for a number of reasons, including:

- the completion of the aforementioned LAPP Phase I & II, NCVA, IAHS and other local climate change risk assessments;
- the release of fundamental data sets, including the Smartline–National Shoreline Geomorphic and Stability Mapping Project, the mid-resolution Coastal Digital Elevation Model and Urban Digital Elevation Model for selected coastal and/or urban centres that will allow the reassessment of specific climate change risks, particularly those related to inundation,[5] and
- the continued domestic and international emphasis on climate change in the lead-up to UNFCCC COP 15 in Copenhagen (December 2009) and its immediate aftermath, linked with the domestic debate on the Carbon Pollution Reduction Scheme (CPRS).

While these events will contribute to prolonging, and perhaps even enhancing, the discussion around climate risk management for adaptation (e.g. particularly if substantial commitments to mitigation do not emerge from Copenhagen linked to real and verifiable implementation mechanisms), discussion in itself does not necessarily translate directly into the optimal use of climate change risk assessment techniques, an expansion of the currently small group of competent practitioners or, most importantly, the implementation of adaptation policies and measures. For example, while demand for risk assessments has grown and adaptation planning is proliferating rapidly, the extent to which such plans are being translated into action remains unclear. To borrow from Herbert Simon (1976), decision makers currently appear focused on what could be described as substantive risk management, which emphasises the identification and quantification of risk and the identification of risk treatment options. Risk management also has critical procedural dimensions associated with defining system boundaries, selecting appropriate stakeholders, the implementation of adaptation policies and measures and the monitoring of outcomes. Marginalisation of such procedural risk management and the failure to articulate a process by which adaptation outcomes will be realised can result in the identification of plausible adaptation responses that sit idle, or the rejection of those responses by interest groups that were not involved in the risk management process.

It is therefore appropriate to review some of the key challenges in translating knowledge about climate risk into adaptation: a) failure to fully understand the problem; b) constraints on institutional capacity, and c) absence of evaluation and monitoring for adaptation. Each of these is discussed in more detail below.

Problem orientation

A common feature of climate change risk assessments and associated adaptation planning is that they fail to examine the problem in its appropriate context. For example, one of the most commonly cited failures of understanding is the historical dominancy of biophysical factors in considerations of climate risk which is, in part, the legacy of early attempts to characterise climate risks (Jones and Preston 2009). Preston and Westaway (Chapter 19, this volume) note that the majority of adaptation plans reflect this assessment paradigm and fail to consider non-climate drivers when undertaking climate risk assessments and adaptation planning, despite the fact that factors such as population growth and demand for environmental amenity are significant contributors to exposure to climate hazards. Such narrow conceptualisations of climate change, risk and its management have the potential to undermine adaptation. As stated by Clark (2002):

> The people in most private organisations and government agencies do not look at decision processes as a whole and as a result often misunderstand why they fail to unfold as hoped. This failure of understanding is usually part of the problem that must be resolved. In many instances, poor decision processes are also the product of badly designed and coordinated policy structures, although they are often misattributed to 'the opposition,' unforeseen events beyond the control of participants, poor public relations, or lack of resources.

To address such bias, the assessment of climate vulnerability and risk has been aligned increasingly with the sustainability science and development communities (Schipper and Pelling 2006; Turner et al. 2006). In so doing, assessments have increasingly attempted to examine both the biophysical and socio-economic dimensions of climate risk in an integrated manner. Almost a decade ago, Geoscience Australia conducted a series of urban risk assessments of natural hazards, which integrated measures of biophysical risk with measures of social vulnerability (Granger et al. 1999; Granger and Hayne 2000; Middelmann and Granger 2000; Jones et al. 2005). More recently, Preston et al. (2008) integrated a diverse array of determinants of climate risk that reflected not only climate hazards, but also the sensitivity of settlements to those hazards and the capacity of households and local governments to manage risk. Such studies can assist in development of more comprehensive, systems-based perspectives on climate risk that may help ensure management efforts focus on key challenges and significant investments are not made in chasing the wrong problem (Barnett et al. 2008).

Ensuring robust problem orientation also requires that risk assessment activities, including the choice, scope and application of assessment tools, are directly linked to policy development and objectives. This means that risk assessment activities should not be conducted in a policy vacuum (i.e. risk assessment without a management context) nor should adaptation decision making be undertaken without a rigorous assessment process (i.e. risk management without a knowledge context). As management responsibilities for human settlements are not confined to local governments alone (Smith *et al.* 2009), implementing such a management framework will necessitate thorough engagement with all three spheres of government – with specific attention to ensuring that local government is fully and properly engaged. Such an engagement must recognise the strengths and weaknesses of each level of government, and an open, transparent dialogue on the best use of their different capacities, experiences and lessons learned to date in climate change impact assessment and adaptation planning. The reality of increasingly complex governance networks, however, also suggests that traditional risk assessment and management approaches that focus on individual institutions may be of limited utility. Ultimately, risk management will require actions and collaboration across institutions, suggesting assessment frameworks and tools are needed that can accommodate such complexity, without being so abstract as to be practically redundant, if they are to support effective on-the-ground adaptation decision making.

Capacity for risk assessment and management

While the concept of adaptive capacity has received much attention in the academic literature (Smit and Wandel 2006; Adger and Vincent 2005; Adger *et al.* 2007; Vincent 2007), its role with specific regard to risk assessment and management has not been investigated in any significant detail. Adaptive capacity, however, is highly relevant as it influences not only the ability of institutions to undertake risk assessments, but also also their ability to implement management responses to address identified risks. The specific capacities required to carry out risk assessments and implement management successfully have different relevance at successive stages (Jones and Preston 2009). For example, the capacities required to carry out the early stages of an assessment are largely technical (e.g. data acquisition and interpretation, modelling), while those required later on become dominated by institutional and governance issues (e.g. risk identification, prioritisation, policy development, and communication). In any case, a broad range of constraints on risk assessment and management for local government have been documented recently (Smith *et al.* 2009).

In Australia, local government is widely recognised as being a key player in adaptation due to its responsibilities for local planning and development approvals. Nevertheless, local governments have consistently identified a range of capacity constraints that impede their ability to simply assess risk, much less implement management solutions. These include access to data regarding climate variability and change, access to expertise to incorporate such data into climate change risk assessments, knowledge of existing assets and infrastructure as well as policy and institutional arrangements that provide disincentives for adaptation. Much of the efforts of the Commonwealth until now, with respect to adaptation, have largely been around building capacity at the local level through, for example, the provisioning of digital elevation models and vulnerability assessment products (see above) as well as funding regional and local assessment studies and adaptation planning through programs such as the LAPP and IAHS. Meanwhile, state and territory governments also are increasingly willing to provide policy guidance to local government to facilitate planning reforms. For example, both the governments of New South Wales and Victoria have recently released guidance regarding what assumptions local governments should make with respect to planning for sea level rise (NSW 2009; VCC 2009). Hence, there are some promising signs that higher levels of government are aligning themselves more closely with the needs of local governments, yet there are still significant policy and capability barriers that have yet to be adequately addressed. The existing risk assessment case studies represent only a small fraction of Australia's local government areas. Scaling up such studies to provide information about risk across the continent and subsequently supporting the implementation of adaptation actions will be a substantial national challenge.

While optimising capacity for risk management and adaptation across (and within) the three spheres of government is one challenge, there is also another challenge associated with risk perception and adaptation at the grassroots level. At present, consideration for climate risk is largely occurring within the context of institutional risk management, in the sense that government institutions are examining climate risks relevant to their own operations and responsibilities, but not necessarily looking further afield in framing the problem. Human settlements are complex systems, however, and as such risk assessment and management activities must reflect such complexity and diversity of spatial and temporal scales. For example, without discounting the role of local and other levels of government in adaptation, it is important to acknowledge that risk is borne not only by institutions, but also by civil society, including households, businesses, NGOs and ultimately, individuals. Furthermore, civil society is likely to be a key player in responding to risk that may scale up to have community or even regional significance.

A number of social vulnerability indices have been developed in an attempt to explore the spatial distribution of disadvantage that may be used to predict populations at particular risk of experiencing adverse outcomes. Cutter *et al.* (2003) and Cutter and Finch (2008), for example, used census data to develop a social vulnerability index (SoVI) that represent the spatial heterogeneity in vulnerability across the United States at a local government scale. In Australia, the Australian Bureau of Statistics' Index of Relative Socio-economic Advantage and Disadvantage represents social vulnerability at the census collection district level (Adhikari 2006). Meanwhile, individual climate assessments have attempted to develop social vulnerability or adaptive capacity indices. Preston *et al.* (2008), for example, used a range of household and local government indicators to map spatial patterns in the adaptive capacity of local governments in the Sydney region. Such studies provide some insight into the capacity of individuals and broader communities to participate in adaptation, although it is questionable whether such indices are relevant predictors of the ability of individuals and communities to adapt. As such, it remains to be seen whether the existing pool of risk assessment studies adequately represent risks or responses at the grassroots level.

Evaluating the success of risk management

Despite the literature on adaptation being full of extensive lists of adaptation options available to various sectors and institutions as well as the aforementioned proliferation of adaptation action plans at a range of geopolitical scales, evidence of anticipatory adaptation actions remains limited (Adger *et al.* 2007). This suggests that either much of the emphasis on risk is limited to the identification and assessment of risk rather than its treatment and/or that the monitoring and evaluation component of risk management has been neglected. Preston *et al.* (2009b) note that much of the work on evaluating adaptation has been confined to the developing world to support adaptation funding programs and projects by multilateral institutions such as the Global Environment Facility to support the requirements for reporting under the United Nations Framework Convention on Climate Change as well as traditional overseas bilateral development assistance (UNDP 2007; Hedger *et al.* 2008). Significant work remains to be done in developing evaluation and monitoring frameworks and relevant metrics that can be used effectively to document the successes and shortcomings of climate risk management efforts.

In Australia, environmental monitoring and reporting has been a common practice across the three levels of government through 'State of the Environment' reporting and other mechanisms. Nevertheless, there is currently little guidance on how such monitoring and reporting mechanisms can be modified to improve the tracking of progress toward adaptation objectives or to monitor key threshold values that may trigger adaptation actions. In addition, Preston and Westaway (Chapter 19, this volume) note that formal adaptation plans often overlook the need for the monitoring and evaluation of prescribed adaptation actions. Addressing this gap should be a priority for all levels of government as monitoring and evaluation are essential to track risk over time, assess the effectiveness of adaptation strategies and facilitate learning and development regarding best practice for climate adaptation.

Research pathways

Risk management is a useful paradigm, but its application to addressing climate risk in Australia is at a crossroads. Some authors have noted the maturation of risk management methods for climate change within the research community (Jones and Preston 2009), which has helped to address past lapses and expanded the toolbox of methods. Yet this crowded toolbox also impedes attempts to incorporate some manner of standardisation into assessment and risk management practice. Therefore, the key research challenge is to enable the transition from the existing suite of case studies/pilot projects into systematised approaches. The underpinning dialectic in such a transition is to maintain the flexibility of method to allow site-specific application, mindful of biophysical, political administrative and capacity contexts, while preserving both rigour and comparability. Of course addressing this creative tension between flexibility and context and rigour and comparability is an underpinning issue for the assessment of climate risk and adaptation planning for all sectors and all places worldwide. It is perhaps most evident in settings where the differential capacity of decision makers (and decision influencers) is high, and the imperative for addressing climate change is also high. These factors appear to apply to the management of urban settlement climate change risk in Australia, demanding a significantly enhanced research engagement with those facing the decision-making challenge that climate change will bring to our settled landscapes.

Despite the challenges, there is evidence that progress is being made in Australia with respect to both adaptation science and policy. Attempts are clearly being made to address key capacity and policy barriers to adaptation at the local scale, although such efforts represent the tip of the iceberg with respect to what is needed. Meanwhile, the emergence of multiple research institutions for adaptation such as the National Climate Change Adaptation Research Facility, the CSIRO's Climate Adaptation Flagship, and the Victorian Centre for Climate Change Adaptation Research have increased the intellectual capability directed at the adaptation challenge and built networks among researchers and other stakeholders. If properly managed, such institutions could assist in developing a coordinated research approach to risk assessment and management in Australia and, through partnership and outreach to stakeholders including adaptation planning practitioners, ensure climate risk is framed in a robust manner and adaptation solutions are identified that reflect the integrated nature of governance in Australia. As a first step toward this goal, however, researchers and decision makers alike must reflect upon past and emerging work on the characterisation of climate risk with a critical eye and learn effective pathways by which adaptation science can effectively support adaptation policy.

Acknowledgements

CZM staff are thanked for their inputs that have helped shape this chapter, in particular Carmen Elrick.

Endnotes

1   <www.climatechange.gov.au/impacts/settlements.html>
2   Disclosure: CZM Pty Ltd is part of the project team for one NCVA.
3   National Coastal Vulnerability Assessment <www.climatechange.gov.au/impacts/coasts.html#research>
4   A full list of successful LAPP Phase I recipients is shown at: <www.climatechange.gov.au/impacts/localgovernment/lapp_round1.html>

Disclosure: CZM Pty Ltd is one of the panel of providers to the LAPP recipients and is leading the completion of three LAPPs and peer reviewing one other.

References

ABS (Australian Bureau of Statistics) (2004) 'Year book Australia – 2004'. Publication 1301.0, Australian Bureau of Statistics, Canberra.

ABS (Australian Bureau of Statistics) (2006) 'Regional population growth, Australia 2004–2005'. Publication 3218.0, Australian Bureau of Statistics, Canberra.

Abuodha PA and Woodroffe CD (2006) 'International assessments of the vulnerability of the coastal zone to climate change, including an Australian perspective'. Report to the Department of Climate Change, Canberra. <http://climatechange.gov.au/archive/impacts/coastal–international.html>

Adger WN, Agrawala S, Mirza MMQ, Conde C, O'Brien K, Pulhin J, Pulwarty R, Smit B and Takahashi K (2007) Assessment of adaptation practices, options, constraints and capacity. In: *Climate Change 2007: Impacts, Adaptation and Vulnerability. Contribution of Working Group II to the Fourth Assessment Report of the Intergovernmental Panel on Climate Change.* (Eds ML Parry, OF Canziani, JP Palutikof, PJ van der Linden and CE Hanson) pp. 717–743. Cambridge University Press, Cambridge, UK.

Adger WN and Vincent K (2005) Uncertainty in adaptive capacity. *Comptus Rendus Geoscience* **337**, 399–410.

Adhikari P (2006) 'Socio-economic indexes for areas: introduction, use and future directions'. Australian Bureau of Statistics, Canberra.

AGO (Australian Greenhouse Office) (2006) 'Climate change impacts and risk management: a guide for business and government'. Department of Environment and Heritage, Commonwealth of Australia.

Barnett J, Lambert S and Fry I (2008) 'The hazards of indicators: insights from the Environmental Vulnerability Index'. *Annals of the Association of American Geographers* **98**, 102–119.

Blackmore KL and Goodwin ID (2009) 'Climate change impacts for the Hunter, lower north coast and central coast region'. Hunter Councils, Inc., Thorton.

Changnon SA, Pielke Jr. RA, Sylves RT and Pulwarty R (2000) Human factors explain the increased losses from weather and climate extremes. *Bulletin of the American Meteorological Society* **81**, 437–442.

Clark TW (2002) *The Policy Process. A Practical Guide for Natural Resource Professionals.* Yale University Press, New Haven.

Cutter SL, Boruff BJ and Shirley WL (2003) Social vulnerability to environmental hazards. *Social Science Quarterly* **84**, 242–261.

Cutter SL and Finch C (2008) Temporal and spatial changes in social vulnerability to natural hazards. *Proceedings of the National Academy of Sciences* **105**, 2301–2306.

CZM Pty (2008) 'Coastal vulnerability and adaptation assessment: a contribution to the compendium of coastal resources tools and methodologies'. CZM Pty, Perth. <www.coastalmanagement.com>

CZM Pty and Damara Ltd (2008) 'Vulnerability of the Cottesloe foreshore to the potential impacts of climate change'. Report prepared for the Town of Cottesloe. <www.cottesloe.wa.gov.au/?p=942>

Granger K and Hayne M (Eds) (2000) 'Natural hazards and the risks they pose to south-east Queensland'. Australian Geological Survey Organisation and the Bureau of Meteorology, Canberra and Melbourne.

Granger K, Jones T, Leiba M and Scott G (Eds) (1999) 'Community risk in Cairns: a multi hazard risk assessment'. Australian Geological Survey Organisation, Canberra.

Hajer M (2003) Policy without polity? Policy analysis and the institutional void. *Policy Sciences* **36**, 175–195.

Hedger MM, Mitchell T, Leavy J, Greeley M, Downie A and Horrocks A (2008) 'Evaluation of adaptation to climate change from a development perspective'. A study commissioned by the GEF Evaluation Office and financed by DFID. Institute of Development Studies, Brighton.

Henderson V (2002) Urbanization in developing countries. *The World Bank Research Observer* **17**, 89–112.

Howe C, Jones RN, Maheepala S and Rhodes B (2005) 'Implications of potential climate change for Melbourne's water resources'. CSIRO and Melbourne Water, Melbourne.

Jones RN (2001) An environmental risk assessment/management framework for climate change impact assessments. *Natural Hazards* **23**, 197–230.

Jones RN and Preston BL (2009) Adaptation and risk management. *Climate Change WIREs*, (submitted).

Jones T, Middelmann M and Corby N (Eds) (2005) 'Natural hazard risk in Perth, Western Australia'. Geoscience Australia, Canberra.

Kirono D, Podger G, Franklin W and Siebert R (2007) Climate change impact on Rous Water supply. *Water* **March**, 68–72.

Mahlman JD (1998) 'Science and non-science concerning human-caused climate warming'. *Annual Reviews of Energy and Environment* **23**, 83–105.

Middelmann M and Granger K (Eds) (2000) 'Community risk in Mackay. a multi-hazard risk assessment'. Australian Geological Survey Organisation, Canberra.

NATCLIM (2007) 'Planning for climate change: a case study'. City of Port Phillip, Melbourne.

Nicholls R (1995) Coastal megacities and climate change. *GeoJournal* **37.3**, 369–379.

NSW (New South Wales State Government) (2009) 'Draft sea-level rise policy statement'. Department of Environment and Climate Change, Sydney.

Pickering C, Good R and Green K (2004) 'Potential effects of global warming on the biota of the Australian Alps'. Australian Greenhouse Office, Canberra.

Preston BL, Brooke C, Smith T, Measham T and Gorddard R (2009a) Igniting change in local government: lessons learned from a bushfire vulnerability assessment. *Mitigation and Adaptation Strategies for Global Change* **14**, 251–283.

Preston BL, Smith T, Brooke C, Gorddard R, Measham T, Withycombe G, McInnes K, Abbs D, Beveridge B and Morrison C (2008) 'Mapping climate change vulnerability in the Sydney Coastal Councils Group'. Prepared for the Sydney Coastal Councils Group by the CSIRO Climate Adaptation Flagship, Melbourne and Canberra.

Preston BL and Westaway RM (2010) A critical look at the state of climate adaptation planning. In: *Managing Climate Change: Papers from the GREENHOUSE 2009 Conference* (Eds I Jubb, P Holper and W Cai) pp. 205–220. CSIRO Publishing, Melbourne.

Preston BL, Westaway R, Dessai S and Smith TF (2009b) Are we adapting to climate change? Research and methods for evaluating progress. *Fourth Symposium on Policy and Socio-Economic Research*, AMS Annual Meeting, 12–15 January, 2009, Phoenix, AZ.

Rappaport J and Sachs JD (2003) The United States as a coastal nation. *Journal of Economic Growth* **8**, 5–46.

Schipper L and Pelling M (2006) Disaster risk, climate change and international development: scope for, and challenges to, integration. *Disasters* **30**, 19–38.

Simon HA (1976) From substantive to procedural rationality. In: *Method and Appraisal in Economics*. (Ed. SJ Latsis) pp. 129–148. Cambridge University Press, Cambridge, UK.

Smit B and Wandel J (2006) Adaptation, adaptive capacity and vulnerability. *Global Environmental Change* **16**, 282–292.

Smith TF, Brooke C, Measham TG, Preston B, Gorddard R, Withycombe G, Beveridge B and Morrison C (2009) 'Case studies of adaptive capacity'. Prepared for the Sydney Coastal Councils Group by the University of the Sunshine Coast and the CSIRO Climate Adaptation Flagship, Sippy Downs and Canberra.

Turner II BL, Kasperson RE, Matson PA, McCarthy JJ, Corell RA, Christensen L, Eckley N, Kasperson JX, Luers A, Martello ML, Polsky C, Pulsipher A and Schiller A (2006) A framework for vulnerability analysis in sustainability science. *Proceedings of the National Academies of Science USA* **100**, 8074–8079.

UN (United Nations) (2006) 'World urbanization prospects: the 2005 revision'. Department of Economic and Social Affairs, Population Division, New York.

UNDP (United Nations Development Programme) (2007) 'Draft monitoring and evaluation framework for adaptation to climate change'. UNDP, New York.

UNFCCC (United Nations Framework Convention on Climate Change) (2008) 'Compendium on methods and tools to evaluate impacts of, vulnerability and adaptation to, climate change'. UNFCCC. <http://unfccc.int/adaptation/nairobi_workprogramme/compendium_on_methods_tools/items/2674.php>

VCC (Victorian Coastal Council) (2008) 'Victorian coastal strategy 2008'. Victorian Coastal Council, East Melbourne.

Vincent K (2007) Uncertainty in adaptive capacity and the importance of scale. *Global Environmental Change* **17**, 12–24.

Willows R and Connell R (2003) 'Climate adaptation: risk, uncertainty, and decision-making'. UKCIP Technical Report, UK Climate Impacts Programme, Oxford.

18. ADAPTING INFRASTRUCTURE FOR CLIMATE CHANGE IMPACTS

Michael Nolan

Abstract

Infrastructure is currently being designed and built based on past climate rather than the future changing climate that it will be exposed to during its expected life. Climate change effects on infrastructure will have a direct financial and reputational impact on corporations and government, especially in circumstances in which assets experience accelerated degradation and fail during extreme climatic events such as storms, floods and heat waves.

Each form of infrastructure (water, power, transport, buildings and communications) has key sensitivities to a change in climate variables, such as extreme wind, solar radiation, extreme rainfall, heat waves and soil moisture. The level of exposure to climate change effects is dependent on the location of infrastructure, its condition and expected service life as well as its dependence on other services, such as water, power and transport.

Maladaption of infrastructure presents significant risk to governments, investors, infrastructure owners and operators.

New investment for infrastructure will need to be 'climate ready' to meet future climate change design compliance. Organisations that are infrastructure-intensive need to understand their existing direct exposure to climate change impacts.

From the author's experience in working on over 20 climate change adaptation projects at AECOM (an international engineering and advisory services consultancy), the following adaptation responses are recommended:

- understand climate risks early to ensure cost–effective adaptation planning for new and existing assets;
- explore climate risks to infrastructure collaboratively between organisations and government jurisdictions related to water, power and transport, and
- use climate change design guidelines for design, materials selection and maintenance regimes.

Introduction

Climate change presents a significant threat to infrastructure in many forms including buildings, coastal developments, water pipelines, transmission lines and road networks. The Australian Bureau of Statistics (ABS 2002), values Australia's homes, commercial buildings, ports and other physical assets at $1957 billion. To accommodate climate change, all nations, states, local government and corporations will need to consider the following:

- vulnerability of existing infrastructure assets and services;
- planning and investment decisions, and
- adjustments to design standards.

Infrastructure has been designed, built and maintained on the premise that future climate will be similar to that experienced in the past. Yet increasing concentrations of atmospheric greenhouse gases are already changing our climate and indications are that they will continue to do so in the years and decades ahead. Planning on the basis of a false assumption presents a significant risk to governments, institutional investors, infrastructure sectors and organisations.

Short-term climate change effects include changes to extreme events of rainfall (flooding), wind, dust storms, cyclones, storm surges, hail, frost, heat waves, bushfires and lightning storms.

Medium-term climate change effects include the accelerated degradation of materials due to greater storms, extreme wind and rainfall intensities as well as increased temperatures increasing corrosion rates. Increased frequency and severity of drought as well as changes to groundwater generates increased ground movement and changes soil chemistry affecting the performance of foundations and structures. Changing climatic conditions are currently not incorporated in the maintenance of the structural integrity of existing assets, let alone new assets.

Longer-term climate change impacts will include sea level rise, coastal inundation and change in salt water intrusion inland. Climate change will affect vegetation, marine and terrestrial ecosystems and landscapes, altering ecosystem benefits for the delivery of infrastructure services.

Climate change adaptation

The international scientific community agrees that the global climate is changing due to the enhanced greenhouse effect. Changes will occur irrespective of efforts to reduce greenhouse gas emissions, although success in achieving mitigation will determine the magnitude and possibly the nature of the changes to which we will need to adapt. It is paramount that nations continue efforts to reduce emissions of greenhouse gases, but alongside such strategies should also be effective, efficient and well-communicated adaptation plans.

Although discussed theoretically since the 1980s, adaptation as an applied response to environmental change only gained practical traction following the Intergovernmental Panel on Climate Change Third Assessment Report in 2001 (IPCC 2001). Adaptation strategies aim to increase the resilience of human and natural systems to possible changes in climatic conditions, whilst taking account of the social dimensions of distributing losses. Such strategies should underpin the frameworks for managing future climate risk, offering the potential to reduce future economic, social, and environmental costs.

The climate change adaptation of infrastructure-related services and assets should be a priority for the sustainability of settlements and their prosperity through the challenges ahead.

Adaptation drivers and benefits

There are several drivers influencing the market to understand and reduce risk to infrastructure associated with climate change impacts including:

- investment decisions that now include climate change as part of investment risk, such as climate change impact considerations for major port infrastructure investment;

- reducing greenhouse gas emissions. Recently the focus has begun to move to adapting for climate change impacts;
- mitigating for climate change effects by early planning, design and operation of infrastructure (for example, see Association of British Insurers information below), and
- all levels of governments perceiving the relationship between risk of infrastructure failure and loss of life from the possible future position of insurer of last resort.

The short- and long-term benefits associated with the implementation of adaptation strategies are noted below.

Short-term benefits

Improved planning for:

- residential, commercial and regional developments, and
- water, power, transport, telecommunications facilities and structures.

Investment decisions generating:

- reduced risk of significant losses or liability, and
- increased confidence in infrastructure projects due to climate change risk mitigation being integrated into project design.

Reputation and public confidence in management associated with:

- government;
- industry, and
- infrastructure assets.

Insurance coverage and reduced premiums for climate change should be assessed and infrastructure projects risk-controlled.

Long-term benefits

- improved resilience of infrastructure to climate change effects;
- reduced asset maintenance costs;
- reduced disruption to services and productivity;
- better informed crisis management, enhancing public confidence and ability to respond to crises, and
- reduced risk to government, industry and/or community possessing no, or inadequate insurance protection.

Adaptation responses to infrastructure risks

The adaptation responses will vary depending upon the level of risk exposure. The adaptation response for any given location, region and infrastructure service will depend on the type and severity of the climate change impact to be avoided, mitigated or managed. The inherent climate and landscape characteristics of a location (i.e. flood- or wind-prone, coastal and marginal rainfall area) combined with potential climate changes, need to inform infrastructure planning, investment, design and operation.

In leading the assessment on the 'Climate Change Impacts to Infrastructure in Australia' as a key input for the Garnaut Climate Change Review (Maunsell Australia 2008), our AECOM team (formally Maunsell Australia) identified and assessed the impacts of climate change on infrastructure in Australia. This study focused in detail on buildings in coastal settlements, water supply infrastructure in major cities, port infrastructure and operations, and electricity distribution and transmission networks. AECOM also assessed roads and bridges, communications and alpine regions.

The results of the assessment highlighted that out of all the economic impacts from climate change, the effects on infrastructure represented about 40% of the total economic impacts to Australia. It is almost certain there will be increased damage to infrastructure from extreme rainfall, wind, bushfire, storm and cyclone events, sea level rise, increased maximum temperatures and ground movement. Water supply issues suffered the greatest impacts early in the 21st century from reduced rainfall, increased temperature and evaporation. The effects are likely to be the greatest in regional areas.

For the infrastructure assets assessed, it was identified that there is expected to be a 2–15% reduction in asset life. Further implications for assets include a likely increase in maintenance, repair and replacement costs and generally lower service reliability.

As part of the assessment, adaptation options were considered for most effects. These adaptation measures were linked to a specific time scale for implementation for each of the seven climate change scenarios being assessed (three business-as-usual and four mitigation emissions scenarios). The key drivers for adaptation were a combination of direct impacts to the infrastructure and its service reliability including reduced life of assets, government policy, planning and regulatory mechanisms such as regulated funding allocation to upgrade asset resilience or adjust maintenance regime, and insurance and investment demands on infrastructure developers, owners and operators.

There are also several barriers to adaptation such as ease of adaptation, knowledge required for adaptation decisions and cost to adapt. Adaptation barriers were modelled to change over time due to regional and global responses to increased climate change impacts. As the impacts from climate change increased, the barriers to adaptation decreased as resources were more likely to be channelled into overcoming these barriers and reducing impact costs.

Risk assessment and adaptation planning

From AECOM's experience, large organisations will benefit from a strategically-aligned climate change risk assessment using ISO 4360 Risk Management (AS/NZS, 4360) processes to identify and prioritise a range of asset and operational risks. Those risks that are identified and analysed should primarily be managed through existing risk management frameworks within the organisation. Where significant and progressive change is required to address risks and opportunities for business development, an adaptation plan is required.

Government jurisdictions (local, state and national) and large corporations should have a whole-of-jurisdiction climate change adaptation plan. While all jurisdictions of government share some responsibility for various aspects of public infrastructure, municipalities carry the greatest responsibility for adapting infrastructure to climate change (Mehdi 2006).

Canada, US, UK, Australia and New Zealand have undertaken assessments of climate change impacts on infrastructure. There have been several studies researching the potential costs. Figure 18.1 shows the likely significance of the cost of adapting to climate change in Canada for a sample of major infrastructure.

The Canadian experience is based on the premise that assessing economic impacts of climate change involves estimating the value of direct and indirect market and non-market costs of implementing adaptation options and the benefits gained as a result of the adaptation (Lemmen 2004).

The Association of British Insurers report (2005) investigated the implications of climate change on the insurance industry. They estimated that without action to restrict climate change, the cost of insured damage in a severe US hurricane season could rise by

Adaption	Estimated cost
Constructing all-weather roads (not on permafrost)	$85,000 per km plus $65,000– $150,000 per bridge
Constructing all-weather roads (on permafrost)	$500,000 per km
Replacing coastal bridges to cope with sea-level rise	$600,000 per bridge
Expanding wastewater treatment capacity (Halifax)	$6.5 billion

Figure 18.1: Estimated costs for adapting selected infrastructure in Canada. Costs are based on a 5% increase in mean temperature and a 10% increase in mean precipitation over the present century. All dollars based on 2001 value in Canadian dollars. (Lemmen 2004)

three-quarters, and that the costs of flooding in Europe could increase the annual flood bill by up to £82 billion. They state that by addressing climate change now, the insurers' increased capital requirements for hurricanes, typhoons and windstorms could be reduced by more than £33 billion, with strong, well-enforced building codes to prevent and reduce windstorm damage. In the UK, effective flood management could save 80% of the costs of flood damage (Association of British Insurers 2005).

In the future, organisations are likely to need to have a climate change plan to maintain existing insurance cover, or maximise the level of insurance cover in the future. It is an obligation, particularly for the corporate and government sectors, to demonstrate that they have taken appropriate (reasonable) steps to avoid an insurable loss. Therefore, insurers are likely to demand that infrastructure owners demonstrate how potential threats of climate change will be mitigated to reduce potential loss exposure.

Land use planning is another important adaptation measure for coastal communities, flood zones and areas facing potential water shortages. For example:

- zoning of residential, commercial and recreational areas will need to be adapted to meet changes in coastal, flood plain and rainfall conditions;
- master and structure planning, development control plans and local environment plans will need to incorporate a strategic approach to climate change impacts in the local council area;
- upgrading the energy efficiency of existing building stock as well as new buildings is required to reduce the demand on energy infrastructure and supply during heat waves;
- planning of essential services and emergency response to build resilience for extreme weather events, and
- power and water planning will need to adapt to manage changed resource demand pressures. Two areas of focus could be:
  - to capture and store stormwater locally to supply the expected shortfall in rainfall and traditional reservoir supplies, and
  - to create local distributed renewable or low carbon power generation to reduce the vulnerability of the electricity supply system during heatwave and extreme storm events.

Due to changes in climate, in future, existing technology may not be able to deliver the services that the community expects. For example, engineering solutions may involve technologies focused on localised stormwater collection systems and local reuse in preference to dam catchment and storage. Future roofing technology for new residential homes may withstand greater wind velocities and storm-related damage, such as water intrusion into eaves. While some technologies will be required to adapt existing buildings to an increased standard of weather proofing.

Innovation of infrastructure design and development will be necessary when climate change coupled with population growth increases resource demand pressures. An example is managing severe water shortages within settlements and throughout agricultural regions that are dependent on irrigation. In some circumstances where the risk to infrastructure is extreme, the withdrawal of services, infrastructure assets and restrictions to development may be required. For example, the impact of sea level rise combined with the increased frequency and intensity of storm surge events may require in some places a removal or relocation of buildings, facilities, water, energy, transport and telecommunications infrastructure.

Design standards, materials selection and maintenance

The guidelines for design standards for infrastructure need to be adjusted to allow for changes in the range of expected extreme events as well as accelerated degradation of materials and structures. For example, in some locations by 2030 in Australia, a 25% increase in the intensity of a 30-minute extreme rainfall event is likely to increase the flood level achieved from the equivalent of a flood that is expected to occur once every 100 years, to instead being likely to occur once every 17 years (Coleman 2003). This is almost a sixfold increase in the frequency of exceptionally large flooding events.

This would also mean that the future height of exceptionally large flooding events would be substantially greater. In this instance, design standards and requirements for flood-prone infrastructure and stormwater management would need to be applied to a greater area and range of infrastructure assets. Determining design thresholds and expectations for infrastructure service will prioritise adaptation responses.

The maintenance regime of assets over time will need to adapt to the acceleration in the degradation of materials and structures. For example, preventative maintenance regimes will need to be applied to buildings, roads, bridges and tunnels to maintain structural integrity and avoid significant effects from climate change impacts on steel, asphalt, concrete joints, foundations, protective cladding, coatings and sealants.

A change in the selection of materials for components that will be exposed to changed conditions will be an important initial step for most forms of infrastructure. The selection should be based on the desired life expectancy of the infrastructure and maintenance regime.

For example, road surfaces in some areas will be exposed to increased maximum temperatures and solar radiation leading to accelerated degradation of the asphalt surface. This will require that the materials mix to resemble that used for regions where temperature and radiation levels are higher.

Similarly, coastal infrastructure such as wharfs, bridges and sea walls will need to cope with changes in sea surface temperature, spray zones and added corrosion. Material degradation in infrastructure may make structures more prone to damage from extreme storm events.

Conclusion

Climate change will affect water, building, power, transport and telecommunication infrastructure assets.

Decision makers in government and private sector organisations need to identify and mitigate climate change risks to infrastructure assets, significantly reducing future financial and social impacts.

Organisations that own and manage infrastructure assets need to understand their direct exposure to climate change impacts. Organisations that are dependent on infrastructure services for business continuity need to understand their indirect exposure to these impacts.

New infrastructure provides an opportunity to incorporate climate change adaptation through new design practices including building in options to adapt later such as a new wharf for a port that has the potential to change in height to adapt to substantial increases in sea level. It is more cost–effective to design in the option to adapt an asset later in its life when the climate change effects are more severe than when it was built. Not designing in adaptation may result in needing to replace an asset completely to meet new and evolving conditions. Alternatively building in all the adaptation options up front may not be economically the best use of resources if the conditions designed for are not likely to be experienced for several decades.

Each infrastructure sector needs to lead the adaptation process with support and guidance from governments to adjust design standards, establish codes of professional practice for incorporating climate change into design, management and maintenance of assets.

Investors in infrastructure need to include climate change resilience in their investment criteria and planning.

References

ABS (Australian Bureau of Statistics) (2002) 'Year book Australia – 2002'. Australian Bureau of Statistics, Canberra.

AS/NZS 4360 (2004) *Risk Management.* Standards Australia/Standards New Zealand.

Association of British Insurers (2005) 'Financial Risks of Climate Change'. Association of British Insurers. <www.abi.org.uk/climatechange>

Coleman T (2003) 'The Impact of Climate Change on Insurance Against Castastophes'. Insurance Australia Group. Presentation to the Institute of Actuaries of Australia. <www.iag.com.au/sustainable/media/presentation-20021219.pdf>

Infrastructure Canada (2006) *Adapting Infrastructure to Climate Change in Canada's Cities and Communities.* Research and Analysis Division, Canada.

IPCC (2001) Climate Change 2001: The Scientific Basis. Contribution of Working Group I to the Third Assessment Report of the Intergovernmental Panel on Climate Change. Cambridge University Press, Cambridge, UK.

Lemmen DS and Warren FJ (2004) 'Climate Change Impacts and Adaptation: A Canadian Perspective'. Natural Resources Canada, Ottawa.

Maunsell Australia (2008) 'Climate Change Impacts to Infrastructure in Australia and CGE Model Inputs'. Report commissioned by the Garnaut Climate Change Review. Melbourne.

Mehdi B (2006) 'Adapting to Climate Change: An Introduction for Canadian Municipalities'. Natural Resources Canada, Canadian Climate Impacts and Adaptation Research Network, Ottawa.

19. A CRITICAL LOOK AT THE STATE OF CLIMATE ADAPTATION PLANNING

Benjamin Preston and Richard Westaway

Abstract

Adaptation is rapidly becoming a mainstream policy response for addressing biophysical and social vulnerabilities to climate change, yet there are few systems in place for tracking progress toward adaptation goals. A general framework and associated criteria for the evaluation of adaptation planning was developed, based upon the existing adaptation planning literature. Applying this framework to a suite of 57 adaptation plans from Australia, the United Kingdom and the United States revealed that many institutions have fallen short in articulating a robust strategy for adapting to climate change. Therefore, while the rapid proliferation of adaptation planning shows that the learning process for addressing climate risk is underway, significant progress will need to be made in the future if adaptation is to be placed on a sound policy foundation. Such progress may be facilitated through expansion of adaptation as both a scientific discipline and a sector of professional practice.

Introduction

Over the past two decades, institutions across a range of geopolitical scales have given significant attention to the identification and implementation of policies and measures to address the risks posed by climate change. Historically, such policy development has been dominated by policies to reduce greenhouse gas emissions through top–down international and national policy initiatives including the United Nations Framework Convention on Climate Change (UNFCCC) and the Kyoto Protocol (UNFCCC 1992; Pielke 1998; Pielke *et al.* 2007); however, recent years have witnessed renewed attention to climate adaptation as a complementary risk management strategy that can progress from the bottom up. This attention can be attributed to a) increased awareness of the vulnerability of human populations to climate variability; b) increasing evidence of an anthropogenic signal in current adverse climatic events, and c) a commitment to dealing with additional climate change and consequences regardless of future emissions trajectories.

The rapid development of adaptation as a mainstream strategy for addressing climate vulnerability is illustrated by a broad range of adaptation policy developments and adaptation planning at different geopolitical scales. For example, at the international level, the UNFCCC has established a Climate Adaptation Fund, currently administered by the World Bank, to fund adaptation programs and projects in the least-developed nations. In addition, the Global Environment Facility has initiated an adaptation pilot program to fund additional climate projects as a foray into adaptation. At national scales, developing nations have completed National

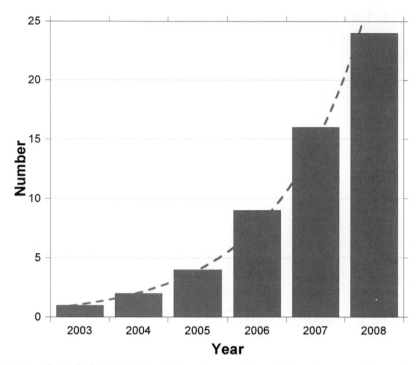

Figure 19.1: Number of adaptation strategies, plans and consultation documents from Australia, the United States and the United Kingdom published over the past six years that were considered in the current study.

Adaptation Programs of Action that represent frameworks for prioritising adaptation needs. Developed nations have also started national adaptation planning. Australia, for example, has developed a National Climate Change Adaptation Framework (COAG 2007) and has made significant investments in adaptation science through the National Climate Change Adaptation Research Facility and the CSIRO's Climate Adaptation Flagship research initiative.

One policy instrument that has seen rapid proliferation over the past six years has been formal adaptation planning, particularly at the state/province and local government/municipal scale. Recently, the number of such adaptation strategies and plans has grown exponentially, doubling approximately every two years (Figure 19.1). Australia, for example, initiated a program in 2008 which provides grants to local governments to assist with climate risk assessment and adaptation planning (DCC 2009). The first round of the program funded 32 projects, with another seven funded under round two. The surge in adaptation planning is also apparent in media reports on climate change (Figure 19.2). Given the activity in this arena, it is appropriate to think critically about the effectiveness of such planning for addressing climate risk. Is adaptation planning simply the policy 'flavour-of-the-month', a worthwhile step on a long-term process of social and institutional learning, or a robust approach to securing near-term reductions in climate risk?

This chapter summarises some of the current efforts regarding the incorporation of evaluation standards and methods into adaptation processes, policies and measures. This is followed by the identification of a plausible set of criteria for evaluating adaptation planning based upon a suite of guidance instruments and an example of how such criteria can be applied to evaluate individual formal adaptation planning schemes. The chapter concludes with the discussion of

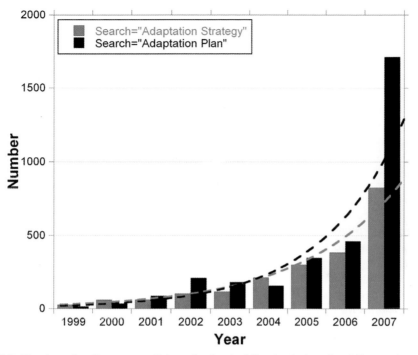

Figure 19.2: Number of online news articles referring to 'climate strategy,' or 'climate change strategy,' or, alternatively to 'climate plan,' 'climate change plan,' 'climate action plan,' or 'climate change action plan.' (Source: Google News archives 2009)

several lessons for adaptation practitioners and the identification of research areas within adaptation science that would be of significant use in supporting future evaluation efforts.

The role of evaluation in adaptation planning

The United Nations has estimated that by 2030, annual investments of US$130 billion will be needed to meet the global demand for adaptation. If the financing and actual implementation of adaptation are to become mainstream components of public policy, then formal frameworks for evaluating adaptation processes and outcomes will be increasingly important. More specifically, there are three key reasons why attention should be given to evaluation.

1. **Ensuring reduction in societal vulnerability** – One of the key goals of climate adaptation is to reduce the vulnerability of human and natural systems to the effects of climate variability and change (or, in other words, the avoidance of 'dangerous' climate change) (UNFCCC 1992; O'Neill and Oppenheimer 2002; Mastrandrea and Schneider 2004). Ensuring that such vulnerability has been reduced requires methods for evaluating and tracking adaptation outcomes. In particular, such evaluation must ensure that the benefits of adaptation policies and measures outweigh the costs (broadly defined) and that additional negative externalities are not created.
2. **Learning and adaptive management** – Climate adaptation is fundamentally a process of social learning. Yet in the absence of methods for evaluating adaptation, opportunities for learning are lost. By tracking the successes and failures of different adaptation initiatives,

institutions can identify effective, efficient and equitable policies and measures. This enables the development of more robust adaptation policy over time in the spirit of adaptive management (Holling 1978).

3. **Need for accountability in an evidence-based policy environment** – From a governance perspective, investments in adaptation and the outcomes they achieve should be transparent. This is true for all aspects of the adaptation process including the development of public communication initiatives, the execution of regional vulnerability or risk assessments, the reform of a given planning policy, or infrastructure upgrades. The development of effective institutional adaptation processes will be assisted by the identification of formal criteria for success, metrics for measuring that success and transparent reporting to stakeholders.

At present, the extent to which adaptation planning is proceeding and succeeding is a matter of polarised opinion. On one hand, some researchers have argued that human beings are inherently adaptive and the history of the species is one characterised by continual adjustment and adaptation to changing conditions and learning about the success and failure of different livelihood and management strategies (Easterling *et al.* 2004). Meanwhile, the Intergovernmental Panel on Climate Change has noted that society is adapting to climate change through both reactive and planned decision making, although it cautions that such behaviour remains limited:

> There are now also examples of adaptation measures being put in place that take into account scenarios of future climate change and associated impacts. This is particularly the case for long-lived infrastructure which may be exposed to climate change impacts over its lifespan or, in cases, where business-as-usual activities would irreversibly constrain future adaptation to the impacts of climate change.
>
> Adger *et al.* 2007

Other researchers have identified significant gaps in adaptation, noting a range of examples where adequate planning has not been conducted for known climate risks in the present day, much less years or decades in the future:

> Despite a half century of climate change that has significantly affected temperature and precipitation patterns and has already had widespread ecological and hydrological impacts, and despite a near certainty that the United States will experience at least as much climate change in the coming decades, just as a result of the current atmospheric concentrations of greenhouse gases, those organizations in the public and private sectors that are most at risk, that are making long-term investments and commitments, and that have the planning, forecasting and institutional capacity to adapt, have not yet done so.
>
> Repetto 2008

Similarly, Easterling *et al.* (2004), while not as critical of the progress on adaptation to-date, cite instances where adaptation actions have not failed entirely, but were not implemented in the most efficient manner possible. As such, they conclude that climate adaptation is likely to progress through a long-term process of 'muddling-through' with occasional winners and losers manifesting on an *ad hoc* basis. Hence there appears to be ample scope for adaptation science and applications to address those areas where adaptation has yet to occur and improve upon the 'muddling through' paradigm to secure efficiency and equity.

Evaluation criteria for adaptation planning

To examine adaptation planning with a critical eye first requires the development of a suite of criteria that represents a benchmark standard for robust planning. While it is tempting to adopt evaluation methods employed elsewhere, these are rare. There has been some discussion among international organisations regarding metrics and approaches for evaluating adaptation (UNDP 2007; Hedger *et al.* 2008). Such discussions largely centre on evaluating substantive outcomes of adaptation actions, but do not necessarily examine the planning processes that generate such actions. Similarly, the OECD has investigated progress on adaptation planning in the national communications made by developing countries to the UNFCCC (Gagnon-Lebrun and Agrawala 2006). The investigation used eight criteria to assess national communications from 39 countries, and distinguished between the process of completing impacts assessments, articulation of intentions to act and evidence of adaptation actions themselves.

At a more local level, Wheeler (2008) analyses a selection of climate change plans that have emerged from state and local governments in the United States. In total, climate change planning was investigated for 64 state and local governments, which involved a review of planning documents produced as well as an interview with at least one official involved in each jurisdiction. It found that plans dealt overwhelmingly with mitigation of, rather than adaptation to, climate change. Only 11 plans mentioned adaptation, and even when they did it was generally as a topic for further research and planning.

Despite the limited scrutiny of adaptation planning in practice, extensive literature on guidance for adaptation planning and policy development exists and continues to grow (Table 19.1). Such guidance embodies current thinking among adaptation researchers with respect to what steps should be taken in adaptation planning. The current study drew upon such literature to assist in defining plausible adaptation criteria. A suite of 20 adaptation guidance documents was reviewed (see Preston *et al.* 2009 for a complete list), with the goal of identifying common concepts and processes that are frequently presented as key components of adaptation planning across a range of guidance instruments.

Through this process, four key stages of adaptation planning were identified (Table 19.2). For each of these stages, multiple adaptation processes were selected, resulting in a total of 19 processes that could serve as relevant evaluation criteria (Table 19.3). The adaptation processes are not termed as 'good' or 'best' practice, as it is recognised that there may be additional concepts that would improve the adaptation process, but which have not typically been included in the adaptation literature to-date. Furthermore, there was significant variation among different guidance documents about which processes to include, reflecting the fact that different guidance documents suggest different methods for adaptation planning. Nevertheless, each of the 19 processes was addressed in one form or another by at least 50% of the guidance documents that were reviewed.

Evaluation of adaptation strategies and plans

The adaptation processes identified in the previous section were subsequently utilised to interrogate a broad range of adaptation strategies and plans that have been produced in recent

Table 19.1 Examples of guidance for adaptation planning

Guidance	Sponsoring organisation
Technical Guidelines for Assessing Climate Change Impacts and Adaptations	Intergovernmental Panel on Climate Change
Guidelines for the Preparation of National Adaptation Programs of Action (NAPA)	United Nations Framework Convention on Climate Change
Handbook on Methods for Climate Change Impact Assessment and Adaptation Strategies	United Nations Environment Program
Adaptation Policy Framework (APF)	United Nations Development Program
Climate Change Impacts & Risk Management: A Guide for Business and Government	Department of Climate Change (Australia)
Adapting to Climate Change: A Queensland Local Government Guide	State of Queensland Government (Australia)
Sustainable Regional and Urban Communities Adapting to Climate Change	Planning Institute of Australia; Department of Climate Change (Australia); Environmental Protection Agency (Queensland, Australia)
Adapting to Climate Change: An Introduction for Canadian Municipalities	Natural Resources Canada
Climate Adaptation: Risk, Uncertainty and Decision Making	Climate Impacts Program (United Kingdom)
Costing the Impacts of Climate Change in the UK	Climate Impacts Program (United Kingdom)
Assessments of Impacts and Adaptations to Climate Change in Multiple Regions and Sectors (AIACC)	Agency for International Development (United States)
Adapting to Climate Variability and Change: A Guidance Manual for Development Planning	Agency for International Development (United States)
Preparing For Climate Change: A Guidebook for Local, Regional, and State Governments	Climate Impacts Group, University of Washington; King County (Washington State); ICLEI (United States)

years. While adaptation planning in developing nations, such as that conducted through the National Adaptation Programs of Action process, has incorporated some degree of evaluation, planning in developed nations has been more *ad hoc*. As such, this study focuses on adaptation planning in a subset of developed nations. At the time of writing, 70 adaptation plans were readily available in the public domain, of which 57 were selected for consideration in the current study. The selected plans were restricted to three countries: Australia (n=18), the United Kingdom (n=20) and the United States (n=19). The plans represent a broad range of geopolitical scales, from local governments to national government agencies. Twelve of the 15 plans addressed a specific sector, such as public health, natural resources management or agriculture, but the majority pursued adaptation planning across a spectrum of issues relevant to the organisation in question.

Adaptation plans also were developed through different processes. For 50 of the 57 (88%), the adaptation plan was prepared within the government or organisation. In many cases, it was undertaken by a single department (often linked to environmental or sustainability functions), but in others it was achieved through the formation of a cross-government working group or committee. The remaining eight adaptation plans were produced by outside bodies, including independent agencies, academic institutions and specialist consultancies.

Table 19.2 Descriptions of four key stages for adaptation planning

Stage	Description
Goal-setting	Establishing what decision makers seek to achieve through adaptation and how performance with respect to obtaining goals will be determined.
Stock-taking	Assessing institutional assets and liabilities that facilitate or hinder adaptation planning and policy implementation. As such, this stage effectively represents an assessment of adaptive capacity; however, to discriminate between different components of adaptive capacity, this stage was conceptualised as assessment of five stocks of capital relevant to adaptation, based upon the sustainable livelihoods literature (Ellis 2000; Nelson *et al.* 2005, 2009).
Decision making	Processes associated with determining what adaptation policies and measures are appropriate. This stage encompasses a variety of tasks including engaging with stakeholders about preferred adaptation responses, assessment of climate and non-climate system drivers, assessment of impacts, vulnerability and risk and the prioritisation of different adaptation options and their harmonisation with existing policy structures.
Implementation and evaluation	Processes associated with the implementation of preferred adaptation options which may include communication, the removal of barriers and the assignment of roles and responsibilities. In addition, this stage also includes processes associated with monitoring and evaluation of implemented actions.

It was also apparent that plans were produced for different purposes. Through the evaluation process, three broad categories of planning document emerged, crudely characterised here as 'strategy documents', 'action plans' and 'consultation papers'. Strategy documents (comprising 30% of planning documents) are primarily strategic in nature and may contain a vision of where a government or organisation wants to be, but provide little in the way of substantive actions for getting there. Consultation papers (14%) outline a planned course of action, but one which is not necessarily intended to be implemented in its current form. They instead invite review or consultation within government itself (where the paper has been written by an independent group) or externally from a wider range of stakeholders, interested parties or the general public. Action plans (54%) provide a structured list of tasks, steps or measures that are planned to be implemented so as to meet a defined adaptation target or goal. Adaptation action plans tend to be functional, typically including details about roles, responsibilities, mechanisms and time scales.

Scoring of adaptation plans

The evaluation of individual adaptation plans was conducted by scoring each plan against the aforementioned adaptation processes. Each process was scored on a three-point scale (0, 1 or 2) resulting in a maximum possible score of 38 (see Perkins *et al.* 2007). The specific requirements associated with each possible score varied among different adaptation processes, but generally followed a consistent system (Table 19.4). Here it may be useful to illustrate the scoring system with a simple example. One of the key adaptation processes is '*assessments of impacts, vulnerability and/or risk.*' For a particular plan to receive a score of 0 for this process, the plan would have to fail to address the implications of climate change impacts, vulnerability and risk for adaptation planning or acknowledge that understanding of such issues may be a core component of planning. A score of 1 would be assigned if the plan acknowledged impacts,

Table 19.3 Descriptions of adaptation processes utilised as evaluation criteria in the current study

Process	Criteria	Description
GOAL-SETTING	Articulation of objectives, goals and priorities	Establishing the objectives, goals and priorities for adaptation.
	Identification of success criteria	Consideration of what successful adaptation will look like and how it will be measured.
STOCK-TAKING	Assessment of human capital	Consideration of the existing skills, knowledge and experience of individuals responsible for adaptation planning and implementation.
	Assessment of social capital	Consideration of the existing governance, institutional and policy contexts for adaptation, including the capacity and entitlements of those institutions, organisations and businesses responsible for designing, delivering and implementing adaptation measures.
	Assessment of natural capital	Consideration of natural resource stocks and environmental services which are sensitive to climate and/or integral in the management of climate risks.
	Assessment of physical capital	Consideration of material culture, assets and infrastructure that are sensitive to climate and/or integral in the management of climate risks.
	Assessment of financial capital	Consideration of stocks and flows of financial resources and obligations within and among individuals and institutions including cash revenue, credit and debt and mechanisms for financial risk management.
DECISION-MAKING IMPLEMENTATION AND EVALUATION	Stakeholder engagement	Engagement of relevant stakeholders and communities throughout the adaptation process.
	Assessment of climate drivers	Consideration of historical climate trends, current climate variability and future climate projections.
	Assessment of non-climate drivers	Consideration of variability and trends in other environmental and socio-economic factors.
	Assessment of impacts, vulnerability and/or risk	Assessment of the impact of changes in climate, vulnerability or resilience to those changes and the relative importance of climate and non-climate risks.

Process	Criteria	Description
DECISION-MAKING IMPLEMENTATION AND EVALUATION	Acknowledgement of assumptions and uncertainties	Transparency about the assumptions made to establish those impacts and risks and the uncertainties involved in their estimation.
	Options appraisal	Identification and comparison of different adaptation options and a means of selecting between them.
	Exploitation of synergies	Identification of where opportunities exist to implement adaptation in a manner that promotes synergies with existing policies or plans, including mitigation.
	Mainstreaming	Identification of ways in which climate change adaptation can be institutionalised or embedded into existing or new policies and plans.
	Communication and outreach	Communication and dissemination of adaptation outputs and outcomes to the appropriate stakeholders and communities.
	Definition of roles and responsibilities	Establishing who is responsible for different aspects of an adaptation strategy.
	Implementation	Establishing the mechanisms that will allow implementation of adaptation measures.
	Monitoring, evaluation and review	Establishing a system of monitoring and evaluation that allows the performance of adaptation to be assessed against success criteria and for review of inputs and procedures.

Table 19.4   Summary of conditions that merit the assignment of different scores to specific adaptation processes

Score	Necessary conditions
0	No evidence of consideration for a particular process was apparent within the published plan. This suggests a particular concept or planning process was neglected.
1	Evidence exists of consideration of a particular process during the development of the adaptation plan. This suggests the concept or process in question was recognised or acknowledged as being of some importance; however, the concept or process remained underdeveloped, suggesting additional consideration may be required for robust planning.
2	Evidence exists of consideration of a particular process during the development of the adaptation plan and significant effort was invested as part of the planning process (or prior to the planning process) to establish a particular process or complete a particular process.

vulnerability and risk as being important but failed to actually undertake or present evidence that some assessment of vulnerability and risk had been conducted (e.g. climate risk assessment was stated as one of the adaptation actions to be undertaken in the future). To receive a score of 2, the plan would have to articulate the various impacts, vulnerabilities and/or risks that are relevant to the planning authority and communicate the process by which such determinations were made.

Regardless of the care invested in articulating conditions by which scores are assigned, this evaluation process is unavoidably subjective. To minimise bias, scores were assigned by one investigator and then reviewed by a second investigator to detect inconsistencies that were subsequently resolved through deliberation. In addition, the sensitivity of the results was tested using an alternative, two-point scoring system that reflected simply the presence or absence of a particular process.

Results

The total scores across all 57 adaptation plans evaluated ranged from six out of 38 (or 16% of the maximum score) to 22 out of 38 (or 61%) (Figure 19.3). This suggests that none of the adaptation plans evaluated provided comprehensive coverage of all adaptation processes. Indeed, only seven adaptation plans scored at least 20 out of 38 (or greater than 50%). Ten adaptation plans scored less than 10 out of 38 (less than 26%).

Further insights can be obtained by analysing the scores based on other factors (Table 19.5). None of the differences observed among different categorisations of adaptation plans were statistically significant, largely due to sample size. There was a slight tendency, however, for longer plans to score higher than shorter plans. Furthermore, the highest score obtained by the shortest adaptation plans (those less than 12 pages) was only 18 out of 38 (or 47%), suggesting brevity provides fewer opportunities to address the evaluation criteria. In addition, 'action plans' tended to score higher than 'consultation papers' and, particularly, 'strategy documents', which again reflects the greater specificity and comprehensiveness associated with those plans designed to guide specific actions.

It is also instructive to examine the adaptation processes which scored highest and lowest across all the adaptation plans evaluated (Table 19.6). For example, those processes where average scores were at least one standard deviation above the mean included, 'assessment of impacts, vulnerability and/or risk', 'communication and outreach', 'articulation of objectives,

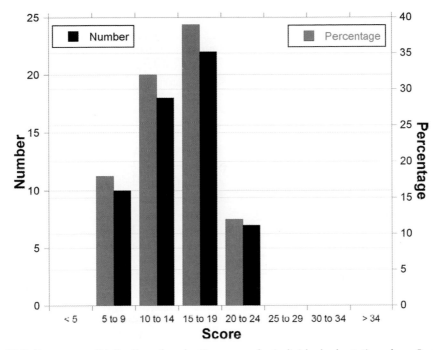

Figure 19.3: Frequency distribution of evaluation scores for individual adaptation plans. Results are presented as either the total number corresponding with each category or the percentage of all adaptation plans falling within a given category.

goals and priorities', 'assessment of climate drivers' and 'implementation'. Even for these processes, however, when a plan received a score greater than 0, in the majority of instances, the assigned score was a 1 rather than a 2. Meanwhile, those processes where average scores were at least one standard deviation below the mean included 'acknowledgement of assumptions and uncertainties,' 'assessment of financial capital', 'assessment of non-climate drivers', 'identification of success criteria' and 'assessment of natural capital'.

These results suggest that adaptation planning is generally following a traditional, and perhaps outdated, approach focused on identifying possible future climate changes and associated impacts, prioritising impacts for intervention and communicating this information to various stakeholders. This approach resembles the Carter *et al.* (1994) guidance for impact assessment, which stressed the acquisition of substantive information regarding consequences, but overlooked the procedural aspects of risk management as well as the interactions between climate and non-climate drivers. Developing an effective strategy for adaptation requires taking stock of existing assets and liabilities. The fact that few documents addressed the financial capacity for adaptation is surprising given that many of the evaluated plans were produced by, or for, government agencies that are traditionally cost-conscious. It is questionable whether informed decisions regarding policy prescriptions for adaptation can be made if the costs and benefits of those prescriptions are not addressed. Furthermore, the failure to adequately consider climate change in the context of other socio-economic drivers that will affect systems of human value creates the potential for biased and incomplete perceptions of risk that lead decision makers down the wrong path. This is exacerbated by the general lack of criteria by which decision makers can define success, raising questions regarding whether decision makers will be able to judge whether or not adaptation has in fact occurred.

Table 19.5   Plan scores stratified by different variables

Stratification variable		Count	Average score	Lowest score	Highest score
Pages on adaptation	< 12	11	13.2	7	18
	12 to 30	17	12.4	6	21
	31 to 60	12	13.4	6	20
	> 60	16	16.4	7	23
% on adaptation	< 20%	4	15.3	9	21
	20 to 95%	31	13.5	6	20
	> 95%	21	14.5	7	23
Country	Australia	18	13.9	8	21
	UK	20	15.0	8	23
	USA	19	13.0	6	21
Geographic scale	National	9	13.1	7	16
	Regional/State	24	14.6	6	21
	Local/Municipal	24	13.7	6	23
Year of publication	2003-2005	7	13.7	9	20
	2006	9	12.7	6	18
	2007	16	13.8	6	23
	2008-2009	25	14.6	7	21
Plan scope	Multi-sector	45	14.2	6	23
	Single sector	12	13.0	7	17
Plan type	Strategy document	17	11.9	6	19
	Consultation paper	8	14.8	8	21
	Action plan	31	15.0	6	23

Notes: Average score represents the average among all planning documents within a given category. Lowest and highest score reflects the score for the worst- and best-performing plan within a given category. Note that the highest possible score is 38.

A number of caveats should be applied to the interpretation of the results. First and foremost, the evaluation methods utilised here assess only information contained within the adaptation plans themselves. Concise, high-level documents may be preferred by public institutions due to their utility in communication. This does not necessarily preclude significant investments of time and resources in adaptation planning behind the scenes. More bottom–up investigative approaches that provide insight into how plans are developed and implemented are needed. Furthermore, the adaptation plans included in this study effectively represent the first generation of plans for which there are few prior precedents or models. As such, what should or should not be addressed in adaptation planning remains a matter for discussion. The fact that existing adaptation guidance varies considerably with respect to recommending steps in planning suggests the greatest utility of adaptation planning at present may be simply learning how to plan better by devising and implementing a plan, i.e. learning by doing. Hence, this study has avoided ranking individual adaptation strategies or plans for performance, as such judgments were deemed of lesser utility than building broader understanding around adaptation planning as a practice where improvements can be made. Finally, it is important to note that while this study focuses on formal declaration of adaptation to climate change, the bulk of adaptation efforts is likely to be informal and may not even be recognised as climate adaptation *per se*.

Table 19.6   Percentage of adaptation plans receiving specified scores for each adaptation process

Deviation	Adaptation process	% of 0s	% >0	% of 1s	% of 2s
>1σ above the mean	Assessment of impacts, vulnerability and/or risk	2	98	74	25
	Communication and outreach	5	95	75	19
	Articulation of objectives, goals and priorities	18	82	65	18
	Assessment of climate drivers	19	81	60	21
	Stakeholder engagement	19	81	67	14
<1σ from the mean	Exploitation of synergies	21	79	67	12
	Definition of roles and responsibilities	23	77	61	16
	Implementation	28	72	61	11
	Assessment of social capital	30	70	53	18
	Mainstreaming	30	70	60	11
	Monitoring, evaluation and review	32	68	56	12
	Assessment of physical capital	46	54	49	5
	Assessment of human capital	56	44	35	9
	Options appraisal	56	44	39	5
>1σ below the mean	Assessment of financial capital	58	42	42	0
	Identification of success criteria	67	33	30	4
	Assessment of non-climate drivers	67	33	28	5
	Acknowledgement of assumptions and uncertainties	68	32	26	5
	Assessment of natural capital	70	30	28	2

Notes: Standard deviations refer to the deviation of the mean score (on a scale from 0 to 2) for each adaptation process across all adaptation plans relative to the scores for all processes.

Conclusions

The current study provides only a high-level, top–down evaluation of adaptation planning, and thus caution must be applied to avoid over-interpretation of the results obtained. Nevertheless, a number of broad conclusions appear, which may provide a starting point for future investigations and assist adaptation practitioners with more robust planning.

- Formal adaptation planning is proceeding rapidly, suggesting adaptation is becoming a mainstream approach to managing climate risk. Its popularity may arise from such plans being a relatively low-cost vehicle for pinpointing relevant actions and focusing institutions on the problem of climate change. While the continual emergence of such planning will fuel public and institutional awareness about adaptation, it is unclear to what extent the current rhetoric of planning will translate into successful adaptation.
- Despite significant activity, if one interprets existing formal adaptation planning at face value, it is clear that the practice of adaptation planning remains underdeveloped. Evaluation of a suite of adaptation plans suggests planning conforms to a system of climate risk management derived from the physical sciences community rather than the more comprehensive and integrated adaptation planning guidance literature.

- There are marked similarities among individual adaptation plans suggesting a process of co-evolution that may result from a small suite of plans acting as models for their successors. This creates the potential for common gaps in planning to be repeated across different institutions, sectors and time periods. It also suggests that examination of the implementation of adaptation plans is necessary to truly discriminate among different initiatives.
- Improving adaptation planning could be achieved through a greater emphasis on goal-setting and stock-taking to ensure a pathway is identified and followed by which adaptation planning can lead to action. In addition, there is clear evidence that adaptation planning could benefit from accounting for socio-economic drivers of change as well as climatic drivers. As such, adaptation should be integrated with other social, urban and regional planning and disaster mitigation efforts, raising the question of whether climate adaptation should remain distinct from other planning processes.
- Efforts should be made to develop more standardised evaluation methodologies for adaptation as a move toward the identification of 'best practice' for adaptation practitioners. To this end, attempts to share learning regarding adaptation planning and build consensus across diverse stakeholders and institutions with respect to the adaptation guidance that already exists would be of greater utility than disparate groups producing additional guidance manuals.
- A range of research pathways may help more robust adaptation planning including the development and maintenance of adaptation databases through which planning and actions can be shared; the identification of appropriate metrics that can be used to evaluate progress and performance toward adaptation, and the longitudinal study of adaptation policy development and implementation (Preston *et al.* 2009).

Acknowledgements

This work was supported through a Julius Career Award granted to the lead author by the CSIRO. The authors also acknowledge the contributions of Dr Suraje Dessai of the University of Exeter and the assistance of William Perkins, US Environmental Protection Agency.

References

Adger WN, Agrawala S, Mirza MMQ, Conde C, O'Brien K, Pulhin J, Pulwarty R, Smit B and Takahashi K (2007) Assessment of adaptation practices, options, constraints and capacity. In: *Climate Change 2007: Impacts, Adaptation and Vulnerability. Contribution of Working Group II to the Fourth Assessment Report of the Intergovernmental Panel on Climate Change.* (Eds ML Parry, OF Canziani, JP Palutikof, PJ van der Linden and CE Hanson), pp. 717–743. Cambridge University Press, Cambridge.

Carter TR, Parry ML, Harasawa H and Nishioka S (1994) 'IPCC technical guidelines for assessing climate change impacts and adaptations'. Department of Geography, University College, London.

COAG (Council of Australian Governments) (2007) 'National climate change adaptation framework'. COAG Secretariat, Department of the Prime Minister and Cabinet, Canberra.

DCC (Department of Climate Change) (2009) Local adaptation pathways program. Commonwealth of Australia. <www.climatechange.gov.au/impacts/localgovernment/lapp.html>

Easterling III WE, Hurd BH and Smith JB (2004) 'Coping with global climate change: the role of adaptation in the United States'. Pew Centre on Global Climate Change, Arlington.

Ellis F (2000) *Rural Livelihoods and Diversity in Developing Countries*. Oxford University Press, Oxford.

Gagnon-Lebrun F and Agrawala S (2006) 'Progress on adaptation to climate change in developed countries: an analysis of broad trends'. ENV/EPOC/GSP(2006)1/FINAL, Organisation for Economic Co-operation and Development, Paris.

Hedger MM, Mitchell T, Leavy J, Greeley M, Downie A and Horrocks A (2008) 'Evaluation of adaptation to climate change from a development perspective'. A study commissioned by the GEF Evaluation Office and financed by DFID. Institute of Development Studies, Brighton.

Holling CS (1978) *Adaptive Environmental Assessment and Management*. John Wiley & Sons, New York.

Mastrandrea MD and Schneider SH (2004) Probabilistic integrated assessment of 'dangerous' climate change. *Science* **304**, 571–575.

Nelson R, Kokic P, Elliston L and King J (2005) Structural adjustment: a vulnerability index for Australian broadacre agriculture. *Australian Commodities* **12**, 171–179.

Nelson R, Kokic P, Crimp S, Howden M, Devoil P, Meinke H, McKeon G and Nidumolu U (2009). 'A preliminary assessment of the vulnerability of Australian agricultural communities to climate variability and change'. CSIRO Climate Adaptation Flagship, Canberra.

O'Neill BC and Oppenheimer M (2002) Dangerous climate impacts and the Kyoto Protocol. *Science* **296**, 1971–1972.

Perkins B, Ojima D and Correll R (2007) 'A survey of climate change adaptation planning'. H. John Heinz III Centre for Science, Economics and the Environment, Washington, DC.

Pielke Jr R (1998) Rethinking the role of adaptation in climate policy. *Global Environmental Change* **8**, 159–170.

Pielke Jr R, Prins G, Rayner S and Sarewitz D (2007) Climate change 2007: lifting the taboo on adaptation. *Nature* **445**, 597–598.

Preston BL, Westaway R, Dessai S and Smith TF (2009) Are we adapting to climate change? Research and methods for evaluating progress. *Fourth Symposium on Policy and Socio-Economic Research*, 12–15 January, 2009, American Meteorological Society, Phoenix, Arizona.

Repetto R (2008) 'The climate crisis and the adaptation myth'. Working Paper 13, Yale School of Forestry and Environmental Studies, New Haven.

UNDP (United Nations Development Program) (2007) 'Draft monitoring and evaluation framework for adaptation to climate change'. United Nations Development Program, New York.

UNFCCC (United Nations Framework Convention on Climate Change) (1992) 'United Nations framework convention on climate change'. United Nations, New York.

Wheeler SM (2008) State and municipal climate change plans. *Journal of the American Planning Association* **74**, 481–496.

Part 3

Communicating climate change

20. RISING ABOVE HOT AIR: A METHOD FOR EXPLORING ATTITUDES TOWARDS ZERO-CARBON LIFESTYLES

Stefan Kaufman

Abstract

This chapter describes and evaluates a method used to explore attitudes about zero-carbon lifestyles. We saw a complex, technical and multi-stakeholder policy debate on an inherently difficult topic to understand, let alone solve, and a gap in the public discussion. What might it be like living in a zero-carbon community? What are the advantages? The disadvantages? We sought to help fill this gap by integrating diverse perspectives across and within community and business to stimulate positive, solutions-focused dialogue. Valuable insights into reactions to some current visions of a zero-carbon future were generated. We found our approach effective in harnessing an existing willingness and capacity for sophisticated, positive discussion on the topic, but saw mixed results in stimulating ownership, ongoing conversation and action. Lessons and possible solutions are presented.

Introduction

Mitigating climate change and adapting to whatever proportion of it that we fail to avoid has profound implications for Australia and the world. To avoid the worst, Australian communities need to reduce their emissions profile to close to zero, and do this while assisting other countries to achieve low or zero emissions themselves. In everything we do and everything we want, energy can no longer be so lightly valued, and a stable climate no longer taken for granted.

At present, the path by which we will mitigate and adapt to climate change is unclear. While the opportunities for increased energy efficiency and alternative energy sources are diverse and promising, and noting the hope that various carbon sequestration technologies may bear fruit, there is no simple technological fix to this challenge. In Victoria at least, a balanced assessment suggests technological options will only get us approximately halfway there (Department of Premier and Cabinet 2007, pp. 124–126). Even were technological options alone sufficient to achieve the necessary emissions reductions, the successful diffusion and adoption of new technologies has complex socio-economic determinants (Rogers 2003). Moving to a carbon reductive civilisation will thus also require societal innovation to an unprecedented degree.

Unfortunately, recent research indicates that while Australians are highly concerned about climate change, this is not matched by in-depth understanding of what mitigation and adaptation mean for lifestyle and future environments. Importantly, people feel they are doing 'enough', and want powerful decision makers in government and business to fix the problem

(Pearce 2008; Department of Environment and Climate Change 2007; Sustainability Victoria 2008). The apparent policy deadlock on the issue seems to indicate that leaders are unsure of what level of social and economic structural changes are acceptable and desirable for the public, particularly in the context of global negotiations. Meanwhile different organised interests have clear, if conflicting, preferences, and are communicating strongly both in the media and behind the scenes.

At GREENHOUSE 2009, Ross Garnaut observed:

> I wondered … whether climate change policy was too hard for rational policy making. It was too complex. The special interests were too numerous, powerful and intense. The timeframes within which effects become evident were too long, and the timeframes within which action must be effected too short.

> Garnaut 2010

Navigating what is feasible and desirable in such a situation is no simple task, but it is far from impossible. Understanding whether or not we can disentangle key aspects such as energy from emissions, energy from economic growth, and economic growth from quality of life are central to achieving a transition. They have broad-reaching implications for all stakeholders. There are no simple answers. Such collective problems call for deliberative, participatory dialogue (Dovers 2005; Funtowicz and Ravetz 1991; Ostrom 2000; Moser and Dilling 2007; Berkhout *et al.* 2003).

The focus of the participatory event described in this chapter was to bring sectoral experts and members of the community together to explore what a zero-carbon Victoria might be like. On 30 September 2008, the Environment Protection Authority of Victoria (EPA Victoria), Monash University and Royal Melbourne Institute of Technology (RMIT) hosted a participatory forum, 'Rising Above Hot Air: A people's inquiry into sustainable lifestyles' (the Inquiry). An associated online forum started on 15 September and ended on 1 November 2008.

We tested the idea that if a diverse group of interested but not highly 'climate-active' citizens could be engaged in solutions-focused dialogue with people representative of key sectors, directions for significant social innovation could be established. Public participants were recruited via two methods: social network recruitment, and random sampling. Participants were screened demographically and for willingness to communicate insights from the workshop. Pre- and post-surveying of participants, post-event evaluation and the insights of the day and online forum around the event were analysed. 'Rising Above Hot Air' suggests some pathways for social innovation, and reflections on effective strategies for generating representative and constructive dialogue on a major collective dilemma. Below are some key design and interpretive considerations, followed by an outline of the event, the insights it generated, and lessons learned.

Challenges in talking about climate change

This section highlights the implications of a body of work on how we think and act in response to risks, and why. An extensive literature on risk, decision making and deliberation from social and cognitive psychology, sociology, political science, anthropology and the emerging field of behavioural economics exists to explain why climate change is inherently a 'wicked' problem for the cognitive and communicative capacities of human beings (see, for example, Kollmuss and Agyeman 2002; Moser and Dilling 2007; Stern 2005; Prendergrast *et al.* 2008; Bazerman 2008; Sandman 2009).

While the detail of these challenges is outside the scope of this discussion, the implications are important for understanding what we learned from our event. Moser (2007) summarises these implications in reviewing literature on risk communication and the risks of climate change. She concludes that threatening information is more likely to be persuasive, cause persistent attitude change and motivate constructive response only when a given audience

- feels personally vulnerable to risk;
- has useful, very targeted information about precautionary action;
- feels it can make a difference (self-efficacy);
- feels recommended actions make a difference (response efficacy);
- believes the barriers to action are low or surmountable;
- docs not value the rewards of inaction more, and
- to deal with all the above, has the time, space and support for wisdom (*sapiens*) to come to the fore over habits and biases (the *'homo'*) that kept us alive on the evolutionary savannah but are barriers now to understanding and managing risks rationally, otherwise perverse and irrational outcomes follow (after Moser, 2007).

This can be treated as a chain of requirements for dealing with climate change. Failing one or more of these points at best leads to the problem being ignored, and at worst to perverse consequences. Moser (2007) mentions the example of someone deciding to buy a large utility vehicle in response to learning of the threat of increases in wild weather due to climate change, instead of reducing their contribution to energy consumption.

The term 'climate change' encompasses many things: the science, mitigation, adaptation and economic adjustment are just some of the divisions that experts on the topic recognise. Even once made, such focused sub-divisions have little direct connection to the lived experience of human beings. The highly personalised nature of effective risk communication highlighted by Moser is thus complicated by the broad nature of the topic when a specific, relevant focus is not carefully maintained. Even once a focus is chosen, noting the possible variations in the cognitive, psychosocial, economic and geographic factors shaping each of the points above for a given individual or group, perceptions and reactions are going to vary across the community. Groups of people reflecting consistent configurations of the factors shaping risk perceptions are sometimes referred to as 'interpretive communities' (Leisorowitz 2007; Kahan *et al.* 2007), and present specific risk communication challenges depending on their current mood (Sandman 2009). Such groups consistently interpret threat information through a particular filter comprising variations of each of the points above.

Taken together, all this means that the same communication is going to be received, ignored or acted on (for better or worse) by different parts of the community based on their particular configuration of interpretive drivers. It suggests that mass communication on climate change needs to be exquisitely crafted if it can be effective at all, and that more two-way participative and deliberative processes may be preferable. Figure 20.1 illustrates this in the form of an engagement matrix that relates the complexity and uncertainty of an issue with the level and nature of engagement required (Robinson 2002). Most aspects of climate change sit in the top right of the chart.

Acknowledging the challenges posed by deeply ingrained human characteristics may seem daunting, even overwhelming, on top of the already well-detailed scientific, institutional and economic challenges that climate change poses. Personally, I find it inspiring. Climate change is caused by humans, with all our foibles, augmenting our actions with complex social and technological systems in a biophysically limited world (Boyden 2004; Doppelt 2008). Reviewing Moser's summary points and the matrix of engagement, we can observe that the civilisation that responds well to the risks of climate change will be one that is worthy of enduring. It

Figure 20.2: Feeling personally vulnerable.

record is further influenced by the decision of which of over 40 cartoons are included. What follows then are cartoons selectively chosen to illustrate the key issues in climate change communication identified above by Moser (2007). Further analysis of the pre- and post-surveys, plus content analysis of audiovisual records of the workshop are the subject of a forthcoming publication.

Content analysis of event records show that issues discussed were wide-ranging, but two elements were most discussed. Firstly, the group was highly engaged with issues of equity, security and access, particularly for the vulnerable, and secondly, very interested and positive about localised lifestyles and sustainable transport options (Vafiadis and Massey 2009).

Figure 20.2 represents a strong theme that emerged at the prospect of increased reliance on public transport. The personal vulnerability triggered by this aspect of climate change, as reported by a number female participants, was a sense of risk and vulnerability, not of climate change, but of personal security risks associated with using public transport.

Figure 20.3: Have useful and very specific response information.

Figure 20.4: Feel THEY can make a difference.

Figure 20.5: Believe anyone's action will make a difference.

Figure 20.6: View the cost of action as acceptable.

Also on transport – participants from outer urban and regional locations were quick to point out that public transport is not an option for them – a blanket instruction to 'do without cars' fails to appreciate their circumstances and options.

Intense interest, and some scepticism, about the prospect of alternative vehicle technologies as an option similarly reflected a sense of dependence on the private sector amongst the group.

A strong sense of curiosity, confusion and scepticism was evident about a plethora of weird and wonderful technologies in the wing for climate change mitigation. An element of 'I'll believe it when I see it' was evident in the group.

Participants were clear that many aspects of quality of life didn't need to be emissions-intensive, but neither did they underestimate the challenge involved in untangling 'necessary'

Figure 20.7: Don't value the rewards of inaction more.

and 'unnecessary' emissions in pursing the good life. Peak oil and its implications for the availability of food, fuel and other energy intensive goods and services was a general concern.

Reflecting their pre- and post-survey results, participants were by and large convinced that action is very necessary and desirable. They were intolerant of any suggestion that a generic solution would work (i.e. community gardens in Broadmeadows, public transport in the country) and somewhat sceptical of the willingness of people not in the room to make major change.

The literature on decision making and risk communication underlines how critical it is that we have the time and support to draw on our social, cognitive and rational faculties to engage constructively with sustainability dilemmas. The quality of the discussion, diverse nature of the participants and high reported levels of satisfaction suggested we achieved this, although imperfectly. No firm conclusions or decisions were made, and follow-up feedback suggests that while participants were happy to continue the discussion with friends and families, they did not extend beyond this to talk to work colleagues, engage in general civil participation or even make further contact with other participants. In part this is because the

Figure 20.8: Time, space and support to draw on their capacitates as a sentient human, not the caveman.

event format relied on use of the internet forum to build follow-on discussion and action, and it was underused. But it also suggests that more time and consideration needs to be given to developing conclusions and making decisions.

Lessons and conclusions

This event highlighted that perceived barriers and opportunities for zero-carbon lifestyles vary across inner-urban, outer-urban and rural-locations and socio-economic conditions. Dialogue touched on issues of access, environment and safety in transport choice, diet and food sources, values around preferred energy sources, and tensions between economic progress and environmental improvement.

It is not intended that anything in this report be interpreted as demonstrating the accept-ability of particular changes, in particular regions, nor at what speed. Indeed, participants recognised that far more could be explored, and some expressed frustration that the format didn't permit doing this.

Community member participants provided further feedback in telephone questionnaires before and after the event, and mid-way in an online evaluation questionnaire. Despite some confusion expressed about the specific goals and desired outcomes of the event, public participants came away with high levels of satisfaction. Six weeks later, there was measurable increase in participants' confidence about their knowledge of climate change. Despite the coincidence of extensive media coverage of the global financial crisis in this period, people entered and left with strong attitudes in support of action on climate change in public, business and private life.

The inquiry format and the leaders worked well, but neither appears to have been critical to the success of the event. Feedback suggests the format may also have detracted from more in-depth participant dialogue and engagement. The online forum was underused but did not appear to impact negatively on satisfaction.

It appears unlikely at the time of writing that the Inquiry had much impact on the broader community and key actors or stakeholders in climate change. Participation from sectoral experts and leaders was less extensive than planned, and media coverage of the event was neither sought nor received. Community members reported a commitment to continuing the discussion with family and friends, but not at work or in civil or political spheres. Participants were willing and able to explore pathways for societal innovation, but need more time and more focused action planning to come to conclusions.

Participants reported valuing each other's contribution more than the leaders' narratives, panel, videos, etc., suggesting a need for better resources from trusted, independent sources; an obvious role for CSIRO and universities.

The experience would be improved for participants in a similar future event if the format was modified to allow greater participation and ownership, to facilitate the discussion towards clear outcomes or conclusions, and if the participants have more involvement in exploring and adding to the goals and outcomes before, during and following the event.

Achieving greater impact beyond the participants of any such event will require concerted action. This should include presenting communications pieces based on the events targeted to specific audiences and mediums, and making attractive invitations to key actors and stakeholders in climate change to join the events themselves. By predicting and responding to how different interpretive communities respond to threat information, these communications can be drafted so as to have greater impact and lead to positive and constructive responses. That said, the diverse and personal reactions to even quite focused discussion points in this event demonstrate the limitations of mass, one-way communication for engaging on this topic. Genuine dialogue and engagement is needed.

Our experience here suggests a need to support many positive discussions in every community, and format may be less important than a positive tone, especially across interpretive communities. Perhaps then, rather than attempting to recruit by demographic representation as we did, different conversations, more focused on agreed outcomes and action, would be desirable, with recruitment based on 'interpretive communities'. As noted previously, these are groups with consistent drivers shaping their interpretation and engagement with a topic. While no such research is currently available in Australia, it should be relatively simple to identify key population characteristics and beliefs of different Australian interpretive communities and recruit on that basis. Alternatively, targeted recruitment via social networks might be desirable, where community leaders or spokespeople known to represent a particular view recruit their communities for a focused conversation (CSIRO EnergyMark project – see Chapter 21 in this volume). The goal in both cases is to understand how different groups in the community consistently hear, interpret and act on information about climate change, in their natural communities. These can then be compared and related across communities, and their self-identi-fied priorities and actions supported systematically.

Overall, 'Rising Above Hot Air' has demonstrated that at least parts of the Australian com-munity are ready and able to have a positive and sophisticated discussion about the implications of a zero-carbon lifestyle. Now many such conversations need to be held, within and between, diverse interpretive communities, with a focus on making decisions and taking actions. All such initiatives will help us become the civilisation that should endure into the future.

This research report was completed under Environment Protection Authority, Victoria con-tribution number EPAVIC 2010-002.

Acknowledgements

The forum concept was created by EPA CEO Mick Bourke, the project team was led by Nicole Hunter, and included Alexa Powell, Jo Sinclair, Sally Jungwirth, Tim Turnbull and myself. Thanks to Dr John Gardner of CSIRO for his invaluable assistance in providing the basis of the assessment questionnaire and general advice.

Thanks are due to the participants, panel members, experts, EPA staff for ideas develop-ment workshops, Monash University, RMIT–ACHRE, Mr Rod Quantock, the Eureka Project (networks, insight and ideas), Long grass Productions (event management) and Head Shift (internet forum facilitation).

Thanks also to Zoe Vafiadis and Rosanne Massey, who conducted a content analysis of topics discussed in the workshop as an assignment in RMIT's Client Based Research Unit of the Bachelor of Social Science (Environment) degree.

References

Bazerman MH (2008) *Barriers to Acting in Time on Energy and Strategies for Overcoming Them.* Harvard Business School, Boston.

Berkhout F, Leach M and Scoones I (2003) *Negotiating Environmental Change: New Perspectives from Social Science.* Edward Elgar, Cheltenham, UK; Northampton, MA.

Boyden SV (2004) *The Biology of Civilisation: Understanding Human Culture as a Force in Nature.* UNSW Press, Sydney.

Department of Environment and Climate Change (2007) 'Who cares about water and climate change in 2007: a survey of NSW people's environmental knowledge, attitudes and behaviours'. Department of Environment and Climate Change, New South Wales, Sydney.

Department of Premier and Cabinet (2007) 'Understanding the potential to reduce Victoria's greenhouse gas emissions'. Victorian Climate Summit 2007 Discussion Papers. Department of Premier and Cabinet (The Nous Group), Melbourne.

Doppelt B (2008) *The Power of Sustainable Thinking: How to Create a Positive Future for the Climate, the Planet, Your Organisation and Your Life.* Earthscan, London.

Dovers SR (2005) Clarifying the imperative of integration research for sustainable environmental management. *Journal of Research Practice* 1(2), M2.

Funtowicz SO and Ravetz JR (1991) A new scientific methodology for global environmental issues. In: *Ecological economics: the science and management of sustainability.* (Ed. R Costanza) pp. xiii, 525. Columbia University Press, New York.

Garnaut R (2010) Climate change and the Great Crash of 2008. In: *Managing Climate Change: Papers from the GREENHOUSE 2009 Conference.* (Eds I Jubb, P Holper and W Cai) pp. 17–28. CSIRO Publishing, Melbourne.

Kahan DM, Braman D, Slovic P, Gastil J and Cohen GL (2007) 'The second national risk and culture study: making sense of – and making progress in – the American culture war of fact'. Cultural Cognition Project at Yale Law School.

Kollmuss A and Agyeman J (2002) Mind the gap: why do people act environmentally and what are the barriers to pro-environmental behavior? *Environmental Education Research* **8**, 239–260.

Leisorowitz A (2007) Communicating the risks of global warming: American risk perceptions, affective images, and interpretive communities. In: *Creating a Climate for Change: Communicating Climate Change and Facilitating Social Change.* (Eds SC Moser and L Dilling) pp. 44–63. Cambridge University Press, Melbourne.

Moser SC (2007) More bad news: the risk of neglecting emotional responses to climate change information. In: *Creating a Climate for Change: Communicating Climate Change and Facilitating Social Change.* (Eds SC Moser and L Dilling) pp. 64–80. Cambridge University Press, Melbourne.

Moser SC and Dilling L (Eds) (2007) *Creating a Climate for Change: Communicating Climate Change and Facilitating Social Change.* Cambridge University Press, Melbourne.

Ostrom E (2000) Collective action and the evolution of social norms. *Journal of Economic Perspectives* **14**, 137–158.

Pearce R (2008) 'Thermometer Survey Report: Our community's response to climate change'. Think Taverner Pty Ltd, Sydney.

Prendergrast J, Foley B, Menne V and Isaac AK (2008) *Creatures of habit: the art of behavioural change.* The Social Market Foundation, London.

Robinson L (2002) 'Consultation: what works'. *Local government public relations conference.* Enabling Change, Wollongong.

Rogers EM (2003) *Diffusion of Innovation.* Free Press, New York.

Sandman PM (2009) Climate Change Risk Communication: The Problem of Psychological Denial. *Peter Sandman Risk Communication website.* New York.

Stern PC (2005) Understanding individual's environmentally significant behaviour. *Environmental Law Reporter.*

Sustainability Victoria (2008) 'Green Light Report: Victorians and the environment in 2008'. Sustainability Victoria and the Department of Sustainability and Environment, Melbourne.

Vafiadis Z and Massey R (2009) Research project: Analysis of an EPA public forum – responses to the prospect of zero carbon lifestyles in Victoria. In: *Unpublished report for Client Based Research Unit, Batchelor of Social Science (Environment).* (Ed. R Lane) RMIT, Melbourne.

21. INVESTIGATING THE EFFECTIVENESS OF ENERGYMARK: CHANGING PUBLIC PERCEPTIONS AND BEHAVIOURS USING A LONGITUDINAL KITCHEN TABLE APPROACH

Anne-Maree Dowd and Peta Ashworth

Abstract

The Commonwealth Scientific and Industrial Research Organisation's (CSIRO) Energy Transformed Flagship is committed to increasing public awareness and knowledge about climate change and the range of energy technologies available to reduce greenhouse gas emissions. In addition to its technical research program, the Energy Transformed Flagship has become increasingly aware of the critical role that changing the individual's energy use will play if we are to be successful in our efforts to mitigate climate change. CSIRO's research has shown that while the majority of Australians are concerned about climate change and greenhouse gas mitigation, they do not necessarily see their own energy use as being part of the problem. The step between concern and action can often be huge, particularly with the presence of conflicting or limited information on the topic and lack of incentives. To address this issue, CSIRO has developed a coordinated initiative called Energymark. This chapter provides evidence for how the Energymark process has tracked changing public perceptions to climate change and energy over time, identified key triggers and barriers and challenges for various individuals to address climate change mitigation at a personal level. Finally, through using the social network analysis method, this research attempts to quantify knowledge transfer through the use of a kitchen table engagement process.

Background to the research

Recently completed surveys conducted for the Centre for Low Emission Technology across Queensland and New South Wales, identified an extraordinarily high proportion (90–93%) of respondents who rate climate change as an issue vital to the nation's future (Ashworth *et al.* 2006; Ashworth and Gardner 2006). The results mirror each other sufficiently to infer that such attitudes apply to the whole of Australia, a conclusion also supported by citizen's panels run by CSIRO in Western Australia, Victoria and New South Wales in 2006 (Littleboy *et al.* 2006). Such a consensus is rare on any highly debated issue, and demonstrates the significant shift in public thinking which has the potential to significantly alter the energy landscape in Australia for the coming decade (NSW Department of Environment, Climate Change and Water 2006; Sustainability Victoria 2008).

Both states' surveys show the Australians best-informed on energy matters are the elderly, the wealthy and the well-educated. Other groups are less well-engaged, especially the 15–24 age group and women. The surveys also reveal there is a strong public appetite for information about new energy choices, broad public interest in all forms of energy, and a declining public tolerance for advocacy of single solutions. Overall the findings are clear; Australians want their energy clean and affordable (Ashworth *et al.* 2006; Ashworth and Gardner 2006).

Despite their concern, many Australians do not necessarily identify their own use of energy as being part of the problem (Ashworth *et al.* 2006). The step between concern and action can often be huge, particularly with the presence of conflicting information and lack of incentives. As such, CSIRO has been grappling with the research question of how to create national momentum around the topic of climate change and its relation to energy that will change the way Australians think and act about energy and climate change mitigation. To address this question, CSIRO has developed a coordinated initiative called Energymark.

Energymark is a concept for brokering public dialogue about the role that individuals can play in moving towards a new energy future. From a theoretical perspective, the central premise of the Energymark process is that individual's values, attitudes and beliefs will drive their intentions, subsequent actions and sustainable changes in behaviour. By understanding individual attitudes and behaviours that can reduce energy consumption, various interventions can be implemented to assist in maximising beneficial social change. The Energymark process is underpinned by social and psychological theories including the 'Theory of Reasoned Action' (Fishbein and Ajzen 1975) and the 'Theory of Planned Behaviour' (Ajzen 1989).

According to Fishbein and Ajzen's (1975) Theory of Reasoned Action, behaviour is a deliberate act which is based on the beliefs of the individual and social norms. Therefore, if an individual has a positive predisposition towards a particular behaviour, and when they perceive support for that behaviour from people around them, then they will form a positive behavioural intention towards that behaviour (Gardner and Ashworth 2008, p. 288). To complement this approach, Ajzen's (1989) Theory of Planned Behaviour recognises that a person's belief about the ease or difficulty in performing a behaviour will be based on their abilities, opportunities and resources (Gardner and Ashworth, 2008, p. 288). For example, if a person feels that they possess the capability and opportunity to change their behaviour, their intention to act will increase, as will the likelihood of actual change.

Taking these two theories into consideration, Energymark is based on the concept that behavioural change requires both new knowledge (to change attitudes) and new policies (to promote action). The process addresses the knowledge contribution stage while at the same time it monitors the effects of policy interventions to bring about large-scale behaviour change (Figure 21.1).

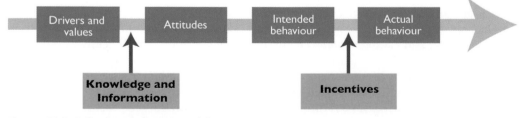

Figure 21.1: Behavioural change model.

The Energymark process

The Energymark concept is based on a three-year behaviour change program developed by the Victorian Women's Trust called Watermark.

> The essence of the Watermark process is that ordinary people bring together small groups of people, meeting monthly to discuss water usage in the home, then sharing their thoughts, anecdotes and 'folk-wisdom', bringing it all back to the Watermark Australia team.
>
> (www.watermarkaustralia.org.au)

The Watermark approach clearly demonstrated the success of utilising pre-existing social networks for generating social change. Because the Watermark groups were formed around friendships and community groups that already existed, feelings of mutual respect, trust and accountability were prevalent and as a result, participants felt safe to discuss the issues and problems openly. Providing new information via the group convenors also helped to build the knowledge base from which to inform any decision-making processes.

Like Watermark, Energymark has found that engaging local community groups in discussions around the kitchen table, and providing them with balanced information and a range of tools, can mobilise them to action to help mitigate the effects of climate change. In addition to providing participants with access to scientific and objective information, CSIRO has included a large focus on measuring and tracking knowledge, attitudes and behaviours in order to provide evidence of the impact and reduction in emissions. This emphasis on embedding monitoring and evaluation processes in a behaviour change process for mitigation is what sets the Energymark project apart from other mitigation behaviour change projects and programs.

The Energymark process brings together small groups of people, meeting at their own pace, to discuss energy technologies and climate change. In the meetings, individuals share their thoughts, anecdotes and first-hand experiences. The benefit of the process is twofold. First, it ensures a coordinated approach to researching public perceptions to climate change and energy technologies across Australia in order to generate insights and provide an empirical benchmark for other researchers. Secondly, engaging the public in this way ensures the information is more likely to be translated into action by individuals because they can relate to the concepts, discuss them openly and change their behaviours accordingly.

Key roles

There are three key roles involved in the Energymark process. These are the Expert Panel, the Secretariat and the Group Convenor (see Figure 21.2).

The Expert Panel is a group of internal CSIRO scientists and external experts who provide Energymark with support by approving all factsheet information, providing answers to technical questions from Energymark participants and safeguarding process legitimacy.

The Secretariat comprises CSIRO representatives who facilitate and evaluate questionnaire completion and responses, provide standardised and balanced responses to Energymark participants' questions, support each group convenor and evaluate their session summary reports.

The Group Convenor role is crucial to the success of the project. Group convenors are the individuals who volunteer to bring together a small group of people (e.g. family, friends, neighbours and workmates) to instigate the process of discussing energy, climate change and the range of mitigation and adaptation options. Group convenors organise and manage the

The Expert Panel
• Defines a standardised topic sequence
• Approves information
• Safeguards process legitimacy

CSIRO (The Secretariat)
• Facilitates questionnaire completion
• Evaluates questionnaire responses
• Provides standardised, balanced information
• Supports the group convenor
• Evaluate convenor responses

The Group Convenor
• Coordinates discussion group
• Faciliates information flow
• Provides a written summary of each discussion

Figure 21.2: The Energymark concept.

meetings of their group and provide the link back to the project Secretariat. They become the conduit for information and at the end of each session send a one-page summary of the discussion results to the CSIRO Secretariat. The Water*mark* project has shown that group convenors are most effective when they have some interest in the topic and may be recruited through a variety of methods including word of mouth, through local interest groups, non government organisations or through advertisements in local newspapers and other media.

Energymark session topics

Methodology

Energymark uses a longitudinal, mixed-methods approach combining quantitative and qualitative methods, including a social network analysis. This allows the researchers to strengthen their knowledge and understanding of how participants perceive the problems of climate change and energy technologies using triangulation of all of the methods, comparing the qualitative and quantitative results in this way increases our confidence in the findings. (e.g. Brewer and Hunter 1989; Creswell 2002; Poole *et al.* 2004; Patton 1990; Tashakkori and Teddlie 1998). Paired sample t-tests were used to compare responses across time. (T-tests compare the means of the two groups of questions at the beginning and the end of the Energymark process with a statistically significant degree of certainty.) Descriptive statistics and open-ended feedback were also used to understand in greater detail participants' experiences.

Research measures

To capture the required measures of the project, a number of survey questions were included. First, beliefs ranging from anti-environmental to pro-environmental were examined. The resulting score is based on the average of 15 items which participants responded '1=strongly disagree' to '7=strongly agree'. Second, environmental versus economic values were tested where participants could answer from '1= the highest priority should be given to economic consideration, even if it hurts the environment'; '2= both the economy and the environment are important, but the economy should come first' '3= the economy and the environment are

Table 21.1 Eight discussion topics and their aims

Sessions and CSIRO supplied resources	
THE BIG PICTURE	
Session 1: Demystifying climate change	Climate change may be due to natural internal processes or external forces, or to persistent anthropogenic changes in the composition of the atmosphere or in land use. The mainstream media has helped raise public awareness of climate change, and most of us now have some understanding of the basic principles of what is occurring. In this session we discuss what is climate change, how we affect it, and how it affects us.
Session 2: Energy and climate change	Energy generation has a particularly close relationship to climate change. The Energy Futures Forum found climate change to be one of the dominant challenges affecting energy in Australia, alongside geopolitical change, innovation and level of community concern about sustainability. Climate change was the most important challenge.
A PORTFOLIO OF SOLUTIONS	
Session 3: New and existing fossil fuel technologies	The third session provides information about coal and natural gas. It also introduces carbon capture and storage. This session provides an opportunity to look at the advantages and disadvantages to new and existing fossil fuel use within Australia.
Sessions 4 & 5: New and existing renewable technologies (parts 1 & 2)	Under Australia's national Mandatory Renewable Energy Target (MRET), all electricity retailers and wholesale buyers have a legal obligation to contribute towards the generation of additional renewable energy – achieved by acquiring a renewable energy certificate. Renewable technologies typically emit low or no greenhouse gases (GHG), and so they assist electricity retailers and wholesale buyers meet MRET obligations set, and when they displace fossil fuels they also help Australia to pursue Kyoto targets for reducing GHG emissions. Both sessions 4 and 5 look at the suite of renewable technologies, including solar, wind, nuclear, biomass, tidal and wave, hydro, and geothermal.
BRINGING IT HOME	
Session 6: Addressing energy and climate change in homes and businesses	Having explained climate change, its relationship to energy and the many approaches to decoupling energy production from GHG emission production, the aim of this session is to apply practices to our homes and businesses, to explore the ways in which we can contribute at these levels.
Session 7: Addressing energy and climate change in the community	By combining many of the technologies and behaviour change measures discussed throughout the sessions, communities can reduce their energy consumption and GHG emissions significantly. The aim of this session will be to explore the role of low emission-distributed energy systems.
Session 8: Final group discussion – addressing transport	An average Australian household generates close to 6 tonnes of greenhouse gas and spends around $8000 each year on transport, of which $2500 is for fuel. Some households spend more than $13 000 each year on transport. This final session provides participants with an opportunity to evaluate their transport alternatives.

equally important', '4= Both the environment and the economy are important but the environment should come first'; and '5= the highest priority should be given to protecting the environment, even if it hurts the economy'.

Third, knowledge was based on participant responses to six questions, where participants answered, 'definitely false' to 'definitely true'. These responses were added in order to allocate a low knowledge to high knowledge score. Participants were also asked to self-rate their knowledge to a range of climate change topics and low emission energy technologies. Fourth, participants' attitudes to the range of climate change and low emission energy technologies were sought and responses could range from '1=strongly disagree' to '7=strongly agree'.

The intentions of participants to undertake more climate and environmentally friendly behaviours were also collected throughout the process. These responses could range from '1=would not accept' to '7=would accept'. Finally, social network data was collected to capture the dissemination of information outside the boundaries of the Energymark groups and to identify key people nodes for potential group convenors.

All the measures used in the project's data-collection phase were developed or amended from previous CSIRO research or scales. Due to the unique topic area and approach, it was difficult to find exact associated measures to capture beliefs, values, knowledge, and attitudes around climate change and mitigation. One limitation to the study is the use of self-rated levels of knowledge. Since there is no existing 'test' to conduct with participants, it was deemed necessary to utilise self-reporting measures. However, the six-statement climate knowledge test does help to verify the accuracy of self-reported knowledge.

Preliminary results

In total, 172 Newcastle citizens had joined the Energymark trial. At the time of writing this chapter, only 30 participants had completed the entire Energymark process and submitted pre- (Time 1), interim- (Time 2) and post-process (Time 3) questionnaires. Therefore, the results reported below reflect the responses of the 30 participants and represents the preliminary results to the trial research. Rates of responding on each survey were typically high (approximately 90% completed), with missing data for each measure reported as appropriate. The results are presented as (1) the demographic and environmental profile of the participants; (2) outcomes of Energymark, including the changes in participants' knowledge, attitudes, behaviours, and social networking, and (3) participant feedback.

Participants' demographic and environmental profile

Participants' ages ranged from 18 to 67 years, with the average age being 37 years. Compared to the Newcastle population, the sample tended to be over-representative of females and people below 39 years. Participants reported a broad range of household sizes, education levels, employment situations and occupations. Specifically, the average household size was three, most commonly participants were in a couple with no children, and the median household income was between $100 000–$124 999. Further, most participants held a bachelor/honours degree, were employed full-time and in professional occupations.

Environmental traits

Environmental traits are characteristics signifying the interest people have in environmental and related issues and what their attitude towards these is. On average the group's responses were slightly skewed towards pro-environmental beliefs, values, knowledge and behaviours.

Table 21.2   Changes in participants' pre-Energymark environmental beliefs, values and knowledge

Measures	Time 2	Time 3
Values	3.68	4.41
Beliefs	4.89	5.24
Knowledge of energy and environment	4.00	4.18

On average, the beliefs, values and knowledge initially reported by the sample Newcastle Energymark participants at Time 1, reported they had moderately pro-environmental beliefs (mean = 4.95 on a 1 to 7 scale). They also valued 'both the environment and the economy' (mean = 3.79). Finally, participants had some knowledge of energy and the environment facts (mean = 4.14).

Participants' environmental traits were followed throughout the process, the average responses for Time 2 and Time 3 are presented in Table 21.2. On completing the final question-naire participants' typically reported stronger pro-environmental values. The mean values reported by the group shifted significantly (statistically large change) from 'the economy and the environment are equally important' to the increasingly pro-environmental, 'both the environment and the economy are important but the environment should come first'. Although the mean pro-environmental beliefs and tested knowledge of the group appeared to be more positive at Time 3, these changes were not substantial enough to be significant.

Knowledge of climate change and low emission technologies

Overall, by the end of the process the group's average had increased to having higher levels of knowledge (5 to 7). Listed in Table 21.3 are the average ratings on climate change topics, nomi-nated before and then after Energymark. The group's average response shifted significantly from moderately low (3) or moderate knowledge levels (4), to higher levels of knowledge (5 to 7). The average change in knowledge of low emission technologies was also significantly positive. At the end of the process participants reported to have higher levels of knowledge (5 to 7). This was also the case for technologies that participants reported to have low knowledge of at the start of Energymark, for example, carbon capture and storage, nuclear and geother-mal. These results are presented in Table 21.4.

Addressing climate change

The group's average responses before and after Energymark are listed in Table 21.5. When measured at the beginning of Energymark, the group was already reporting agreement with

Table 21.3   Changes in knowledge of climate change

How would you rate your knowledge of the following?	Time 1	Time 3
Climate change	4.91	6.50
Greenhouse gas emissions	4.82	6.82
Government initiatives to reduce greenhouse gas emissions	4.09	5.32
Electricity conservation in the home	4.91	6.86
Industry initiatives to reduce greenhouse gas emissions	3.14	5.32
Electricity conservation in the workplace	3.82	6.50
Increasing the price of electricity to reduce greenhouse gas emissions	3.64	6.36

Table 21.4 Changes in knowledge of low emission technologies

How would you rate your knowledge of the following?	Time 1	Time 3
Wind	5.05	6.32
Carbon capture and storage	2.77	6.36
Nuclear	3.59	6.32
Hydro-electric	3.73	6.50
Coal	5.23	6.91
Natural gas	4.73	6.64
Geothermal (hot rocks)	3.77	6.18
Solar	5.82	6.73
Biofuels	4.09	6.73
Oil	4.91	6.73
Wave/tidal	3.55	6.14

the statements. Throughout the process these average responses were reported to be significantly more positive, and by the end, the average was as high as 'mostly agree' (6) or 'strongly agree' (7).

Increased action – doing more to address climate change

Throughout the process participants were asked to identify the types of environmental and climate-friendly behaviours they engaged in. At the interim stage of the process (Time 2), participants reported they had been carrying out a range of environmental behaviours except less accessible initiatives such as increasing the use of ethanol in the car and subscribing to green energy. By the end of the process, nearly all participants were engaging in a range of environmental behaviours and considered their household energy efficiency to be a bit more efficient than similar households.

Listed in Table 21.6 is a comparison between the behaviours people reported doing at the beginning of Energymark and those at the end of the process. By the end all of the participants, except for one non-respondent, were reported to be engaged in a range of behaviours.

Table 21.5 Changes in attitudes toward addressing climate change

How strongly do you agree with the following?	Time 1	Time 3
Climate change is an important issue for Australia	6.25	7.00
The production of electricity is a major contributor to greenhouse gas emissions	6.07	7.00
Industry should be doing more to reduce greenhouse gas emissions	6.36	7.00
People should be doing more to promote electricity conservation in the home	6.50	7.00
Government should be doing more to reduce greenhouse gas emissions	6.50	7.00
People should be doing more to promote electricity conservation in the workplace	6.54	7.00
Increasing the price of electricity to help reduce greenhouse gas emissions	5.54	6.82

Table 21.6   Change in environmental behaviour

Circled 'yes'	Time 2	Time 3
I pay extra for green electricity	33%	97%
I recycle my garbage	90%	97%
I use pesticides in my garden[1]	10%	10%
I use public transport when possible	73%	97%
I carpool	50%	97%
I deliberately buy organic food products	40%	97%
I consider energy efficiency ratings when purchasing white goods	93%	97%
I use plastic bags when shopping[1]	67%	58%
I have a solar hot water system in my home	93%	97%
I have donated money to environmental groups	53%	97%
I use low energy light bulbs	90%	97%
I have signed petitions relating to environmental issues	50%	97%

1: These are behaviours that have a high environmental impact.

The changes in behaviours seemed to be prompted by a combination of increased knowledge through the reading and understanding of the Energymark materials and the small group process. The Energymark process integrates the use of group support mechanisms provided by the existing relationships between members (e.g. sharing of ideas, encouragement, and trusted discussions) and peer pressure (e.g. accountability for change, and change in group norms). This is supported by qualitative comments made on the final survey:

> 'I could not have done this on my own. I received great support from family and friends also doing Energymark with me.'
> 'I could not get away with just saying I would do something. I actually had to do it because otherwise I would be found out, they come to my house all the time.'

Social networking

It was found that by the end of the Energymark process participants, on average, communicated with 34 people about the information provided and the project (see Figure 21.3). By using the social network analysis approach we have been able to not only quantify the spread of information but also identify key active people nodes to which we can approach and ask for their interest in being a Group Convenor.

Qualitative responses

Throughout the course of the trial, a vast amount of qualitative data was collected from Group Convenor discussion reports and feedback, open-ended survey questions, and individual submissions of opinions and remarks through emails and letters. A brief summary of some qualitative data collected to-date can be found below in Table 21.7. For this chapter, a preliminary list of barriers and challenges facing participants has been extracted in order to provide an overview of the key areas preventing behaviour change within the community.

Barriers to behaviour change

Throughout the Energymark process, several barriers and challenges to behaviour change were identified. The first of these was personal or cultural disposition. At the beginning of the

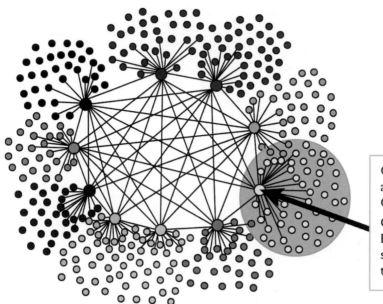

Can target this node as a possible Group Convenor.

Over the course of Energymark they have spoken to 45 external to the group.

Figure 21.3: Social Network Graph Time 3.

Energymark process it was felt that a small number of individuals did not care about climate change or the amount of energy they use, and this was linked to existing attitudes and experiences as well as upbringing. It was also felt that perhaps culturally, it may not be considered as important to modify one's energy usage. The next barrier identified was political. Participants emphasised that they were limited by the incentives and regulations set by the Australian Government in changing their behaviour.

Another key barrier to behaviour change was individual living situations. Participants reflected a sense of disempowerment due to the level of influence on energy consumption, waste recycling and household purchasing. Participants commented that they can be limited by the role they have in a household; for example, if a guest in someone's home, rent/board, share housing.

Economic cost to changing behaviours was also identified as a key barrier to behaviour change. Participants noted that they could not afford the investment required to change to more energy-efficient processes. For example, solar or wind power and even conversion to gas from electricity were viewed as an unrealistic expense. In addition, inadequate information was also highlighted as a barrier. Participants stated that they simply did not have the knowledge or access to information required to understand the importance or relevance of alternative solutions available. Finally, physical and structural factors were identified as a barrier to change. Individuals are largely dependent on the basic infrastructure around them. If the infrastructure of their building or neighbourhood does not provide the capacity for alternatives then the residents are limited in their ability to adapt.

Discussion and conclusion

In summary, the Energymark project has many outcomes and benefits in addressing the issue of behaviour change in Australia. Since this chapter has been written, the trial of Energymark

Table 21.7   List of other behaviours that changed as a result of attending the Energymark discussions

Please list any other behaviour that you changed as a result of attending the Energymark discussions
• (1) less beef consumption (2) less car use (3) more awareness of carbon footprint of goods to point of sale.
• changed all the light bulbs in the house to compact fluorescent lightbulbs CFLs.
• eating less beef and watching my electricity consumption.
• haven't changed as I was doing this stuff already.
• investigating solar and decreasing electricity use.
• looking into solar panels on my home.
• using public transport more often.
• my overall attitude has changed through this process – thank you for opening my eyes to this important issue.
• recycling more, using public transport more, changed light bulbs, conserving energy by reducing electricity.
• solar power use within house.
• using public transport more often and buying organic produce.
• using the train more and monitoring my waste/recycling.

has extended across five Australian states. There are currently 72 Group Convenors with 1092 participants in Energymark groups. In addition to creating momentum at a national level around the topic of climate change and energy, Energymark provides enormous opportunities for the range of stakeholders working in the energy-related area. These include:

- the ability to bring about large-scale behavioural change in Australian energy consumption patterns;
- the ability to track changes in public perceptions to climate change and energy debate as it progresses;
- the impact of information on public perceptions, which is a pronounced measurable take-up of change practices,
- new technologies and approaches at an individual household and community level;
- a way to engage a broader group of stakeholders more readily in the debate;
- a way to involve individuals and communities in the decision-making process around energy;
- information for policy makers, industry and technology developers to inform their areas of expertise, and
- a way to provide financial knowledge and solutions to individual energy and climate change impacts.

In conclusion, Energymark empowers members of the public to discuss energy-related issues with their own peers and within the comfort of their own networks. This increases the quality of the dialogue and its ability to impact on individuals. It provides the opportunity for CSIRO to aggregate, analyse and disseminate results and information from potentially a huge range of participant communities.

The balance and credibility of the information provided through the Energymark process is critical to ensuring a legitimate debate occurs as well as building confidence that participants are part of an unbiased move towards the future. In this respect, the constitution of the Expert Panel and the design of participant questionnaires are all important.

Energymark provides the opportunity to demonstrate leadership in the brokering of knowledge about climate change, energy and mitigation options, thus setting the scene for the effective implementation of Australian policy in this field. To-date, the project has provided many stakeholders (e.g. government, non-government, environmental groups, community associations, industry bodies and general public) with empirical evidence of its impact and has provided the opportunity to present CSIRO's ability to provide detailed findings at varying levels of analysis in relation to this critical topic.

References

Ajzen I (1989) Attitude structure and behavior. In: *Attitude Structure and Function.* (Eds AR Pratkanis, SJ Beckler and AG Greenwald) pp. 241–274. Erlbaum, Hillsdale, NJ.

Ashworth P and Gardner J (2006) *Understanding and Incorporating Stakeholder Perspectives to Low Emission Technologies in New South Wales.* Centre for Low Emission Technology, Pullenvale, Australia.

Ashworth P, Pisarski A and Littleboy A (2006) *Understanding and Incorporating Stakeholder Perspectives to Low Emission Technologies in Queensland.* Centre for Low Emission Technology, Pullenvale, Australia.

Brewer J and Hunter A (1989) *Multimethod Research: A Synthesis of Styles.* Sage Publications, Inc., Newbury Park, CA.

Creswell J (2002) *Educational Research: Planning, Conducting, and Evaluating Quantitative and Qualitative Research.* Pearson Education, Upper Saddle River, NJ.

Fishbein M and Ajzen I (1975) *Belief, Attitude, Intention and Behaviour: An Introduction to Theory and Research.* Addison-Wesley, Reading, MA.

Gardner J and Ashworth P (2008) Towards the intelligent grid: a review of the literature. In: *Urban Energy Transition: From Fossil Fuels to Renewable Power.* (Ed. P Droege) pp. 283–308. Elsevier, Oxford, UK.

Littleboy A, Boughen, N, Niemeyer S and Fisher K (2006) 'Societal Uptake of Alternative Energy Futures, P2006/784'. CSIRO, Pullenvale, Australia.

NSW Department of Environment, Climate Change and Water (2006) 'Who cares about the environment in 2006?' The New South Wales Government, Sydney <www.environment.nsw.gov.au/community/whocares 2006.htm>

Patton M (1990) *Qualitative Evaluation and Research Methods* (Second Edition). Sage, Newbury Park, CA.

Poole MS, Hollingshead AB, McGrath JE, Moreland RL and Rohrbaugh J (2004) Interdisciplinary perspectives on small groups. *Small Group Research* **35**, 3–16.

Sustainability Victoria (2008) 'Green Light Report: Victorians and the Environment in 2008'. The Victorian Government, Melbourne. <www.greenlightreport.sustainability.vic.gov.au>

Tashakkor A and Teddlie C (1998) *Mixed Methodology: Combining Qualitative and Quantitative Approaches.* Sage Publications, Thousand Oaks, CA.

Wilcox D (1999) *The Guide to Effective Participation.* Brighton, Partnership Books.

22. TALKING CLIMATE CHANGE WITH THE BUSH

Clare Mullen, Shoni Maguire, Neil Plummer, David Jones and Colin Creighton

Abstract

Climate variability and climate change are critical issues for Australia, particularly for those in agriculture and rural communities. What do those involved with agricultural industries want to know about the changing climate? How can that information best be communicated?

This chapter will describe how the 'Communicating Climate Change to Agricultural Industries' project addressed these two core questions using a multidisciplinary, consultative approach. The emphasis is on content and contributions from the Bureau of Meteorology (Bureau), given that most of the authors work for the Bureau. A background section highlighting the need to provide climate change information for decision making starts the chapter. This is followed by a description of the methodology used in this project to address the two central themes noted above. Results are presented and discussed. This is followed with some concluding remarks to highlight the key ideas related to our two core questions.

Background

Ongoing rainfall deficiencies in south-west Western Australia since 1970 and in south-east Australia since 1996 have sharpened interest about our future climate, particularly with respect to future rainfall. Figure 22.1 shows the post-1950 drying trend over eastern and southern Australia, some of the main agricultural regions of the country.

Recent droughts in Australia have not been the only cause of a dramatic jump in the interest in climate change (Power *et al.* 2008). The release of Al Gore's movie *An Inconvenient Truth* in 2006 increased world interest and awareness in climate change. The UK Stern Review on the Economics of Climate Change was published in 2007 (Stern 2007), followed by the Fourth Assessment Report by the Intergovernmental Panel for Climate Change (IPCC 2007). Articles about climate change as monitored across 50 world newspapers jumped in late 2006 (see Figure 22.2), with increased interest maintained for much of 2007 (Boykoff and Mansfield 2009).

The IPCC report in 2007 was the fourth report in the series published over a 20 year period, but the first to note conclusively that the recent warming of the climate system was unequivocal and very likely due to the observed increase in anthropogenic greenhouse gas concentrations. Finally, in November 2007, the updated CSIRO and Bureau of Meteorology climate change projections (CSIRO and Bureau of Meteorology 2007) for 2030 and 2050 for Australia were released, receiving mass media attention with publication via most major newspapers and TV stations. Those in climate-sensitive industries, such as agriculture, were particularly interested to find out more about climate change.

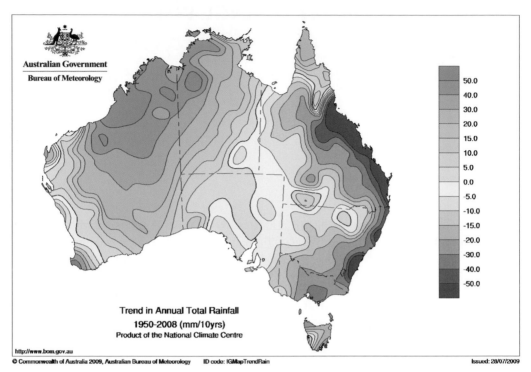

Figure 22.1: Trend in annual total rainfall, 1950–2008. (Source: www.bom.gov.au/cgi-bin/climate/change/trendmaps.cgi?map=rain&area=aus&season=0112&period=1950)

To meet this heightened interest, four related project proposals from separate groups, all aimed at assisting farmers to respond to climate variability and change, were later merged to form the 'Communicating Climate Change to Agricultural Industries' project. Funding came from the Natural Heritage Trust II (NHTII) funds, with project delivery during 2008. The aim of the project was to improve the understanding of the implications of climate change for farm businesses and the environment.

Method

This section describes the methodology used in this project and discusses the reasoning behind some of the choices made. It begins by outlining the organisations involved, the regions where the workshops and forums took place, the format and content of these events, then finishes with a description of how the workshops and forums were evaluated.

Organisations and expertise

The Managing Climate Variability (MCV) program provided overall leadership and coordination for the project.

The Bureau of Meteorology acted as the key science provider of climate change information. Information on regional weather drivers as well as climate change background, observations and projections were provided.

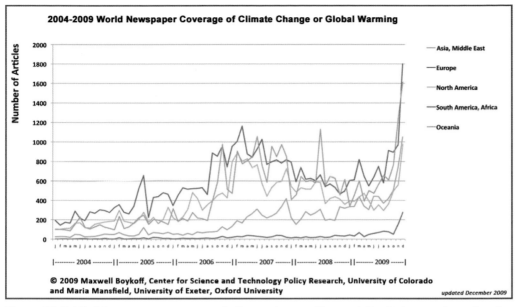

Figure 22.2: World newspaper coverage of climate change. (Source: http://sciencepolicy.colorado. edu/media-coverage/)

Four other project partners provided the following expertise, namely

- the Bureau of Rural Sciences (BRS) presented information on projections of production levels with changing agronomic conditions;
- Meat & Livestock Australia (MLA) focused on emerging international carbon trading systems, as a precursor to the release of a discussion paper on Australia's proposed carbon trading scheme;
- the Birchip Cropping Group (BCG) acted as the representative producer group with a focus on the grains industry. They coordinated workshop logistics with other local user groups. BCG also sourced local agribusiness experts for delivery of regionally and commodity-specific information modules covering marketing, adaptation, farm viability and the social nature of decision making, and
- Econnect Communication provided communication support to team members and the project in general. Their role was to identify repetition and inconsistencies in the information, as well as editing content (e.g. information sheets) for general readability as appropriate.

Focus regions

The pilot areas selected for delivery of the workshops and forums were all mixed farming (grain/livestock) regions: the Wimmera in Victoria (Vic), the north-east agricultural region in Western Australia (WA), and the Eyre Peninsula in South Australia (SA).

Both the Wimmera (Vic) and the north-east agricultural region (WA) were already experiencing prolonged drought. The Eyre Peninsula (SA), where rainfall is comparatively reliable, was also experiencing dry conditions.

Workshop formats

A consultative approach to developing the content of the forums was used. This consisted of two main stages.

The first stage of the project was a program to support key people from grower groups, state agencies, natural resource management organisations, private sector and agribusiness to work with leading experts on climate change to gain knowledge and work towards becoming 'Masters of Climate'. The program delivered regionally specific information, coordinated communication materials and tools for management, with support from scientists and organisations involved in climate change. The 'Masters of Climate' Workshops lasted two days, and were held in May and June 2008. Using the expertise of attendees at the Masters workshops and their knowledge of regional knowledge needs, modules and presentations were then adapted and focused for shorter, sharper delivery to farmers and community – the second stage.

The second stage of the project was a set of Farmer Forums, with information provided on climate change effects, mitigation, adaptation and opportunities at the regional and farm scale for a more general audience. The Farmer Forums lasted half a day, and were held in July and August 2008.

Course content

Previously, weather and climate variability and change information delivered by the Bureau of Meteorology to agricultural groups would generally consist of a talk covering recent weather conditions, the season ahead, climate change science and projections. The content of the presentations would be put together by scientists after a short consultation (usually a telephone call or email) with a representative of the agricultural group.

This project could go beyond answering the 'has our climate changed?' question, as could be answered by Bureau expertise. Using the multidisciplinary approach, the next question of 'what do I do now?' could then start to be addressed for some of the many affected areas of rural industry life. Topics included yield and adaptation simulations, mitigation options, marketing projections, farm viability, and social impacts on climate adaptation.

The Bureau of Meteorology prepared four information modules on the topics of weather drivers, climate change background information, observed climate change and projected climate change.

Formats included factsheets and PowerPoint presentations. Factsheets were later reviewed and converted to pdf documents for the web.

Evaluations

Project evaluation was undertaken from a few different perspectives. Econnect reviewed presentations as the workshops evolved, and provided informal feedback to presenters.

BCG commissioned Roberts Evaluation Pty Ltd to evaluate the learning experience of those involved in the workshops and the forums, and the level of increased knowledge of participants. Roberts observed a Masters workshop and a Farmer Forum in Birchip, Victoria. Three types of questionnaires were also assessed: a pre- and post-workshop questionnaire, an end-of-forum questionnaire, as well as a questionnaire of organisational staff (BCG). The results are referred to as the 'Roberts Evaluation'.

BRS also conducted their own evaluations for an additional report. Surveys were filled out by participants at Farmer Forums, herein referred to as the 'BRS evaluation'. In total, 106 responses were obtained from 240 attendees – a respectable response rate of 44%.

All evaluations are internal documents to the project, but results are reported.

Table 22.1 Project events and participant numbers

Place	Date	Participants
Workshops		
Birchip (Vic)	27/28 May 2008	23
Geraldton (WA)	3/4 June 2008	18
Waite (SA)	25/26 June 2008	26
Forums		
Birchip (Vic)	3 July 2008	400–500*
Keith (SA)	21 August 2008	Day 1 50 Day 2 80
Geraldton (WA)	29 August 2008	20

* Approximation only as part of a larger event and participants changed throughout the day

Results

The following section provides a summary of the number of participants at each of the forums, the mix of the participants and also presents some of the key results from the three different evaluations that were conducted during the course of the project.

Table 22.1 lists the presentation locations, dates, and number of participants.

For the Masters workshops, attendees comprised a mix of local farmers, advisors and consultants. Each received a package containing up to around 20 separate information sheets. Over each two-day workshop, a series of presentations was made by project partners and local experts.

Using feedback from Masters participants, the project team reviewed presentations and developed Farmer Forum agendas for each region. Feedback from those who attended both the Masters workshop and the Farmer Forums indicated that the content was much more locally focused and relevant for the forum. Farmer Forums were intended to coincide with existing events to maximise attendance; however, this was only feasible for the Birchip Farmer Forum, which was conducted in concert with the BCG Grains Research Expo.

There was substantial demand from workshop participants for immediate access to information sheets and presentations; such was the interest in climate forecasting, climate risk management and climate change. The information was later made available on the Birchip Cropping Group website for subsequent reference (www.bcg.org.au/cb_pages/Communicating_Climate_Change.php).

Feedback from formal and informal evaluations suggested that of the Bureau content, the weather and climate drivers presentation was the most popular, particularly the examples of particular weather events and their impacts. Much discussion ensued from the 'Australian Climate Influences' slide (Figure 22.3), indicating the level of interest from producers in understanding the mechanisms producing rainfall in their local area better.

Documented rainfall declines in the focus regions, particularly in Victoria and South Australia (since 1996) and south-west WA (since 1970) meant that many participants have already experienced rainfall declines similar to those suggested by future climate projections. Rainfall projections were relative to the 1980–1999 period; however, few received their actual average rainfall during this period. If climate projections suggested a 10% decrease in rainfall for 2030, for example, this amount was potentially more than the actual rainfall received at the end of the 20th century, hence an improvement of their rainfall situation. Farmers tried hard to understand if their recent climate would be their future reality or just a 'dry run', particularly

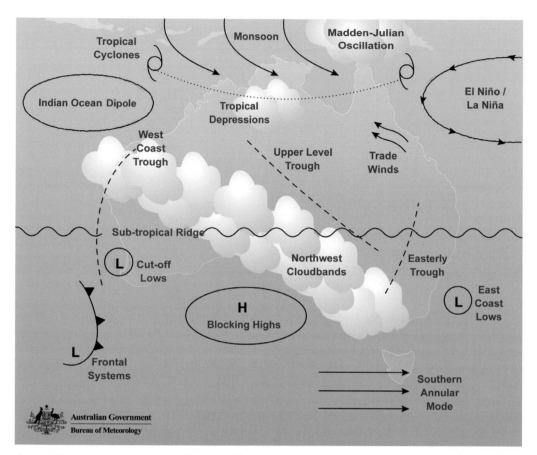

Figure 22.3: Australian climate influences. (Source: www.bom.gov.au/watl/about-weather-and-climate/australian-climate-influences.html)

in South Australia and in Victoria. Sadly, climate science does not provide a clear answer to this question at the present time. This sort of feedback has implications not only for future workshops and events, but also for future research and delivery of climate information at a broader level.

In the BRS evaluation, farmers expressed a strong demand for more information about climate change, including drivers of future climate patterns. They emphasised a need for information that is local, practical and can be applied at farm level. The top three information channels in order of preference were workshops, training sessions, and the internet.

Participants were asked which time frames in the future they would like climate information for, and allowed multiple votes. Farmers showed a clear preference for information on climate change for 10 to 20 years hence (BRS evaluation). There is little existing climate science that addresses this time frame, but it is a clear priority time frame for agricultural industries.

From the Roberts evaluation, respondents' future requests were evenly spread, wanting more on forecasts and models (20%), mitigation and adaptation for farmers (17%), also carbon trading (15%). For those topics that could be addressed by Bureau expertise, future presentations would include information on forecast accuracy, future climatic extremes, as well as changes in weather drivers in Victoria (such as 'blocking highs').

Discussion

The multidisciplinary approach, with expertise in different areas from different organisations, was valuable and successful for both participants and presenters. Discussion moved from weather drivers to climate change to yield impacts and adaptation options, then on to mitigation options and marketing projections, and then bringing all this together to discuss what it might mean in terms of farm viability and social impacts. Running workshops with other organisations meant that the discussion could move well beyond observed and future (physical) climate changes. It allowed the information to be presented in a manner that was more aligned to the operations of the farmers, thus making the information much more relevant and digestible. It also emphasised how much adaptation farmers have already done to survive the last decade.

From the Bureau's perspective, the workshops were successful both for gaining direct feedback on climate and weather-based products, as well as getting the message across to farmers that there are ways and means of reducing the risks from climate variability. The Bureau found it useful to answer direct questions from the audience and give more detailed explanations, particularly to those unsure or sceptical about climate change. The informal discussions were sometimes as useful as the formal presentations, both as a way of learning about farmers concerns as well as ensuring the messages being delivered were clear.

Feedback was also received on presentation skills. Presenters were reminded not to bombard people with too many graphs and diagrams, to explain content better (e.g. axes on graphs) and to perhaps cover fewer topics in more detail, rather than speeding through content. In the evaluations conducted, participants in the Masters workshops rated those presenters who provided time for interaction and discussion during the presentation very highly. Some other presentations ran over time and cut into this valuable discussion time.

Throughout each of the Masters workshops, participants made it very clear that information to be presented in the Farmer Forums needed to be locally relevant. The project team found that gathering local intelligence through discussions with participants and consultants/researchers in the regions was crucial for ensuring this local focus. Widening the participation to engage the stakeholders and capture local knowledge was paramount to the success of this project. This also enabled local examples to be used to demonstrate adaptation/mitigation opportunities and local situations to be superimposed on to modelling and projections of markets and commodities.

The response of participants to the climate science presentations provided valuable insight into user needs regarding long-term climate projections and 'seasonal' predictions. Participants indicated that there was a more urgent need for shorter-term predictions (e.g. seasonal and the next decade) that are more relevant to their decision-making time scales than projections of climate in 2030 or 2070. Despite the focus on climate change, there was much interest in outlooks for the following and subsequent seasons.

Many informal discussions started with 'how much rainfall have you had; how much rainfall will I get in the next few days/weeks/months?' A future project would do well to address all time frames in future presentations better, including the short- and medium-term outlooks, as well as future climate change projections for 2030. The preferred focus would be on the next 10 years, as participants generally saw this as their realistic planning horizon, rather than a future climate 20 or more years away. This will require a restructuring of the way the science is presented.

Greater inclusion of local experts would also improve workshop content, as participants reminded us of the importance of local content and knowledge. Scenario planning in the workshop or development of case studies of how people are improving on-farm management given the current rainfall decline would also be useful. This format may be more effective than the serial 'science–impacts–responses' approach where the focus is first and foremost on the decision-making needs and where, for example, the science and effects are integrated in a way that best informs the scenarios or case studies. This would increase complexity in workshop planning but would place climate information as one of several considerations in making decisions within a risk management context.

Some media coverage was achieved in this project; however, more promotion through local media would be certain to enhance the spread of information. Development of a special media kit, and/or running additional events whilst in the local area would help spread information about climate change.

Workshops should also address resilience, i.e. how to cope with the climate change already experienced so as to help participants to deal with climate change in the near future. Outcomes would be focused on creating 'Masters of Adaptation and Resilience', by addressing current business weaknesses and enhancing knowledge about future climate change and variability.

This project trained 67 'Masters of Climate' across three key agricultural regions in 2008. In excess of 550 landholders, agribusiness, landcare, agency and producer group staff across southern and western Australia have improved their climate knowledge through participation at the four Farmer Forums.

Conclusion

Agricultural industries want climate change information to be presented in a manner that is more aligned to their operations, thus making the information much more relevant and digestible. This means that climate science needs to address the farmers' preferred time frame of the next 10–20 years, as opposed to the current climate change projections provided for 2030 and 2070 (20–60 years ahead).

Those involved with agricultural industries want information on a wide variety of topics that affect their businesses. These include yield and adaptation simulations, mitigation options, marketing projections, farm viability and social impacts on climate adaptation. We found that a multidisciplinary approach was the most effective way of addressing the full range of participant questions and concerns.

The use of local knowledge held by local experts to shape the content of presentations, handouts and other information improved the usability and take-up of information and is recommended wherever possible.

Acknowledgements

On behalf of the participating farmers, agribusiness consultants and agency personnel, we would like to thank the Department of Agriculture, Fisheries and Forestry for providing the opportunity to conduct this successful and timely project communicating climate change.

The consortium comprised:

- Bureau of Meteorology;
- Bureau of Rural Sciences;
- Meat and Livestock Australia;
- Birchip Cropping Group;

- Econnect Communication, and
- Land & Water Australia's Managing Climate Variability.

The team was managed by Colin Creighton from Managing Climate Variability.

References

Boykoff M and Mansfield M (2009) *2004–2009 World Newspaper Coverage of Climate Change or Global Warming.* Environmental Change Institute, University of Oxford. <www.eci.ox.ac.uk/research/climate/mediacoverage.php>

CSIRO and Bureau of Meteorology (2007) 'Climate change in Australia – technical report 2007'. CSIRO and Bureau of Meteorology, Australia. <www.climatechangeinaustralia.gov.au>

IPCC (2007) Climate Change 2007: Synthesis Report. Contribution of Working Groups I, II and III to the Fourth Assessment Report of the Intergovernmental Panel on Climate Change. IPCC, Geneva, Switzerland.

Power S, Plummer N, Pearce K, Walland D, Edwards S, Jones D, Gipton S, Holper P and Whitehead R (2008) Changes in Australian attitudes towards global warming. *MeteoWorld*, April 2008. Geneva. <www.wmo.int/pages/publications/meteoworld/archive/april08/australia_en.html>

Stern N (2007) 'Stern Review on the Economics of Climate Change. Executive Summary'. HM Treasury, London. <www.hm-treasury.gov.uk/sternreview_index.htm>

23. USING GOOGLE EARTH TO VISUALISE CLIMATE CHANGE SCENARIOS IN SOUTH-WEST VICTORIA

Christopher Pettit, Jean-Philippe Aurambout, Falak Sheth, Victor Sposito, Garry O'Leary and Richard Eckard

Abstract

Climate change has been defined as a 'diabolical' policy problem facing society today (Garnaut 2008). Many streams of climate, environmental and social sciences are required to provide evidence-based advice to communities and decision makers alike in dealing with this multi-faceted problem. Significant research has been undertaken by research organisations such as CSIRO and Monash University to improve on modelling the implications of the Intergovernmental Panel on Climate Change (IPCC) climate change scenarios. How are the findings of these often complex scientific outputs communicated to bring about the necessary behavioural change by communities, planners and policy makers? Geo-visualisation of climate change data, models and scenarios is one technique to assist in communicating our climate change projections better. This chapter will present some innovative ways to communicate climate change futures through the use of spatial technologies such as the Google Earth digital globe and 3D rendering and animation software. This will be done in the context of a multi-disciplinary research project being undertaken in south-west Victoria.

Introduction

Climate change is acknowledged as one of the single greatest threats to global socio-economic well-being. Climate change poses significant threats to existing urban infrastructure, current water-use practices and agricultural industries, to name a few (IPCC 2007). In the State of Victoria, under the auspices of the Victorian Climate Change Adaptation Program (VCCAP), research into enhanced communication and participatory technologies is underway. This research is focusing on geographical visualisation (geo-visualisation) technologies, which are being developed and tested as a front-end to climate change data, information and models in south-west Victoria.

Geo-visualisation enables outcomes of social, economic and environmental analysis to be brought together using visual media to convey meaning to land managers, communities, industry, regional planners and policy makers. Geo-visualisation provides a powerful toolkit moving beyond traditional flat maps and creating virtual landscapes where users can collaboratively explore past, present and future climate change scenarios (Pettit *et al.* 2006). To-date there is a limited body of work in applying geo-visualisation technologies for communicating climate change scenarios (Sheppard 2005; Dockerty *et al.* 2005). The VCCAP climate change visualisation research endeavours to contribute to this international research and work towards better tools to enhance the understanding of climate change effects and adaptation options.

A range of geo-visualisation technologies including digital globes (such as Google Earth), 3D scene- and movie-rendering packages (3ds Max, Visual Nature Studio) are being deployed to create a number of geo-visualisation products to enable a better understanding of the complex array of scientific data and models being created to inform policy-makers and land managers on climate change adaptation strategies. Such geo-visualisation tools can be used to build plausible alternative futures, also known as *what if?* scenarios (Klosterman 1999) or *futurescapes* (Lovett 2005).

So far a number of geo-visualisation products have been developed representing land use data and land use impact modelling for climate change scenarios up until 2050 in the south-west Victoria pilot region. These interfaces include geo-visualisations of reduced size climate change models and a number of land use impacts models. Geo-visualisation has also been used to develop a 'farmscape' of the future, which illustrates a farming system which has adopted innovative land management practices in order to adapt to the likely effects of climate change. In this chapter we will present a number of geo-visualisation communications outputs that have been developed by the VCCAP research team.

Geo-visualisation: an enhanced climate change communication tool

Geo-visualisation provides a geographic context to scientific information (data and models) through the use of maps and spatial viewers. Such information can be communicated in two dimensions (flat maps), three dimensions (including elevation and object extrusion) and four dimensions (time). Geo-visualisation is an emerging field which draws upon approaches from many disciplines including cartography, scientific visualisation, image analysis, information visualisation, exploratory data analysis and geographical information science (Dykes *et al.* 2005). Geo-visualisation is an effective communication medium for both community and public policy engagement. It also provides a common language to support information exchange between different science disciplines. Therefore, when dealing with the multifaceted problem of climate change, geo-visualisation provides an extremely valuable tool. Geo-visualisation provides a medium where different researchers can visually communicate and share their scientific results across a range of disciplines and also communicate results to communities, industry and public policy-makers. Figure 23.1 illustrates a geo-visualisation approach for

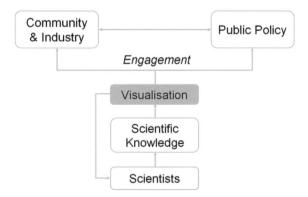

Figure 23.1: Geo-visualisation: an innovative medium for communicating climate change. (Adapted from Pettit *et al.* 2006)

enhancing the communication of geographical phenomena to relevant stakeholders in dealing with climate change adaptation. Geo-visualisation is one of the communications tools used to engage policy makers and regional stakeholders in presenting likely climate change impact and adaptation options in south-west Victoria.

South-west Victoria case study

While research on climate change impact and adaptation is a high priority in most agricultural regions in Australia, the VCCAP needed to pilot its methodology at a regional scale initially. South-west Victoria was selected as the pilot region, based on the following key criteria.

- An agricultural industry production worth over $2 billion annually.
- The region includes a wide range of farming systems including dairy, grains, broadacre livestock grazing, forestry and fisheries.
- The region contains a mix of irrigation and dryland systems.
- A clear rainfall gradient from north to south, but also a clear drying signal in rainfall patterns, consistent with climate change predictions (CSIRO and Bureau of Meteorology 2007).
- The region is also undergoing enterprise and land use change driven by both climate change and commodity prices.

In addition to these, a new community action group, the South West Climate Change Forum, provided VCCAP with a regional stakeholder reference group that represented all major agricultural industries in the region.

The south-west region was defined as the Glenelg Hopkins and Corangamite Catchment Management Authority boundaries, being the coastal strip not more than 200 kilometres wide, from the South Australian border to Melbourne.

Visualising climate change trends and adaptation futures

Substantial research has been undertaken into modelling and understanding climate change and its subsequent effect on natural and human systems. There is a growing body of research into understanding climate change impact, adaptation options, and required institutional adaptation and societal behaviour change (Kelly and Adger 2000). A more recent challenge is the communication of the complex science underpinning our climate change impact and adaptation knowledge in a way that it is understood beyond the science community including land managers, communities, industry and policy makers. Geo-visualisation using digital programs such as Google Earth is becoming more prevalent in helping to map and communicate climate change effects. For example, Google Earth was used to assist in mapping the disaster response of Hurricane Katrina (Nourbakhsh *et al.* 2006). In this research we are applying digital globe technology to overlay the outputs of climate change models, land use models and to create 360° panoramas of current and future farming systems. When mapped in this way, these farming systems are portrayed as current and future 'farmscapes'.

Communicating the downscaling of climate change models

The challenge is to resize global climate change models down (currently at a resolution of around 200 km^2) to a scale more usable for regional adaptation, especially by planners and communities, to improve our understanding of potential climate change effects at the regional/

local level. In this research, we applied the state-of-the-art North American Regional Climate Change Assessment Program (NARCCAP) protocol (see www.narccap.ucar.edu/) and selected the Weather Research and Forecasting (WRF) model for regional downscaling (Skamarock *et al.* 2005). The topography and land use data inputs into the downscaled climate change model provided by the US Geological Survey (USGS) are much more detailed than the conventionally sized global climate change model. At this stage, however, the regional downscaling model was not systematically calibrated against weather observations.

Three emissions scenarios were used as input data for downscaling: SRES B1 (low global warming, specify equivalent or atmospheric carbon dioxide concentrations about 550 ppm by 2100), SRES A1B (medium-range global warming) and SRES A2 (high global warming, carbon dioxide concentrations about 820 ppm by 2100) (IPCC 2007). The global climate model simulations carried out by using the Community Climate System Model and based on the emission scenarios were applied as initial and boundary conditions for the downscaling. The Monash e-research computing facility was used to downscale the scenarios from 200 km (Figure 23.2) down to 60 km, then down to 20 km and finally to a 6.7 km resolution (Figure 23.3). Downscaled climate data were tentatively validated by comparing annual means and variability at two weather station locations situated in Hamilton and Terang, and a first order comparison have been undertaken with the conformal cubic atmospheric model (CCAM) downscaling output.

Associated with the challenge of actually downscaling climate change models is the challenge of making this information accessible, and conveying to regional planners and policy-makers what a downscaled climate change model is, and what the results look like. This is where geo-visualisation can assist. Figures 23.2 and 23.3 illustrate how geo-visualisations were used to display increasing levels of spatial resolution, from a 200 km resolution (Figure 23.2) to 6.7 km (Figure 23.3). The concept of downscaling was illustrated through the use of a grid mesh, generated for specific cell size within a geographical information system (GIS), and

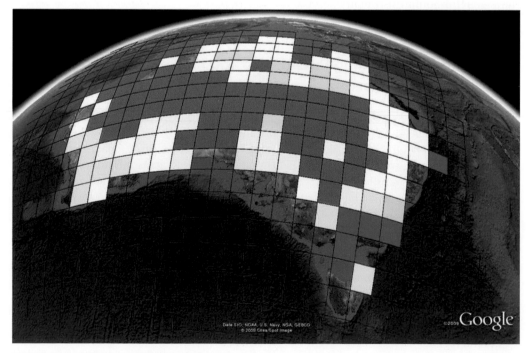

Figure 23.2: 200 km resolution GCM.

Figure 23.3: 6.7 km downscaled GCM.

overlaid on a land cover grid of identical resolution. The use of a grid mesh of various cell sizes (appearing in pink in Figures 23.2 and 23.3) allows the illustration of how the number of cells computed by the climate change model increases with the level of spatial resolution. The use of land cover maps illustrates the gain in spatial resolution associated with the use of smaller cell size as the 6.7 km land cover data (Figure 23.3). The resolution of downscaled climate change data provides a much closer match to on the ground conditions than the 200 km dataset (Figure 23.2).

It is emphasised that the purpose of these geo-visualisations was to illustrate the importance and relevance of downscaled climate data to regional and local government decision makers. In our engagement with planners and policy makers, it is clear that many do not fully understand the concept and importance of downscaled climate change data for appreciating implications at the regional and local scale. Geo-visualisation provides a useful approach in being able to communicate the importance and relevance of downscaled climate change data for better decision making. Further research and development is currently underway with the project team and its collaborators in finalising an authoritative set of downscaled climate change data for Victoria.

Visualisation: a front-end to land use models

Future planning, be it at regional or farm level, is becoming increasingly complex, factoring in the impacts of climate change and what communities, industry and government need to do to adapt. We therefore need to use predictive climate and land use models to help us understand feasible futures and, at the same time, we need innovative ways to communicate and visualise likely futures to support decision making.

Several regional models have been developed and run within south-west Victoria to investigate the likely effects of climate change on natural resources and agricultural production. In this section we will briefly outline some of the preliminary results of the objective and bio-physical process-based models including land suitability analysis (LSA) (Sposito *et al.* 2008), the Agricultural Production Simulator (APSIM, www.apsim.info) and the Sustainable Grazing Systems (SGS) pasture growth model (Johnson *et al.* 2003), and we illustrate how this information can be communicated through the use of geo-visualisation technologies such as digital globes, 3D rendering and animation packages.

Visualising land suitability analysis

Land suitability can be defined as a measure of how well the qualities of a parcel of land match the requirements of a particular type of land use (FAO 1976). The process of determining land suitability for various agricultural commodities takes into consideration topographical, bio-physical characteristic and expert metrics. Land suitability analysis (LSA) is an objective mod-elling approach which applies multicriteria evaluation (MCE) in a Geographical Information System (GIS). The MCE approach applied in south-west Victoria is based on Saaty's (2000) analytical hierarchy process (AHP). The AHP method has been applied to assess land suitability for eight agricultural commodities relevant to south-west Victoria given current and future climatic conditions. The crops belong to three groups: grains (barley, oats and winter wheat), pasture (lucerne, phalaris and ryegrass/sub-clover) and forestry (blue gum and radiata pine). A LSA model was developed for each of those crops using inputs determined through a series of stakeholder workshops with participation of growers, regional/local resource planners regional and experts in agronomy, soil science, climate science and geography.

For example, a LSA model has been developed for wheat (existing variety) – which as a commodity contributed approximately $60 million to the region's agricultural productivity in 2007 – was developed to produce a yield of 7–8 tonnes per hectare per year (ton/ha/year) with common agriculture practice. In Figure 23.4, the areas with the higher suitability are shown in darker green colours. A value of 100% (in the legend) suggests that the yield in those areas will be around 7–8 ton/ha/year; in areas where the suitability is shown as 70%, the yield would be lower, about 5.5 ton/ha/year. A slight decrease of around 4% in terms of the suitability of the land for wheat (not taking into account eminent increases of carbon dioxide – termed 'fertilising effect') would occur. With a reasonable degree of uncertainty factored in the modelling, the overall change in productivity would be negligible.

The final maps of LSA initially developed in a GIS environment, were exported into the Keyhole Markup Language (KML) to create a map overlay that could be displayed in Google Earth (Figure 23.4). The advantages of this format conversion are multiple as (i) KML data can viewed in a digital globe that allows a large degree of interactivity and contextual information that can enhance the value of the displayed dataset; (ii) data in KML format can be easily shared with stakeholders without them having to purchase expensive GIS software, and (iii) the KML format has become an industry standard recognised by the open geospatial consortium (OGC). KML can therefore be viewed via a number of freely available and proprietary spatial browsers, allowing the end user to view the dataset in the viewer of his/her choice.`

Visualising the results of the APSIM model

In addition to the LSA modelling was the application by DPI researchers of predictive wheat modelling from the Agricultural Production Simulator (APSIM) (Keating *et al.* 2003), calibrated for the Hamilton area (see Figure 23.5). For a high global warming scenario, with the effect of increased carbon dioxide and a fertiliser application of around 100 kg nitrogen per hectare factored in, the result shows a median yield increase of about 3%. This is similar to the result from the regional scale LSA modelling, previously discussed.

Figure 23.4: Regional land use suitability for wheat year 2030 (IPCC A2 scenario) in south-west Victoria.

Comparing this result to the Wimmera-Mallee Region, which is likely to have a 20% decrease in wheat crop yield, the south-west Region looks potentially more promising from a

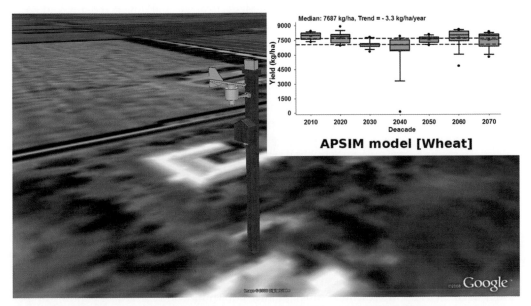

Figure 23.5: APSIM modelling results calibrated to the Hamilton weather station.

wheat productivity perspective. However, it must be emphasised that these are preliminary results and further research is required to calibrate these modelling outputs. The role of geo-visualisation is to provide a geographical context to these preliminary modelling results and provide another way of communicating results and obtaining feedback from key stakeholders within the region; and by policy-makers.

Climate change future farmscapes

Change in rainfall, temperatures, solar radiation, and the number of frost days are some of the main climate change variables which impact agricultural systems significantly. Farming communities are used to dealing with changing climatic conditions on a seasonal basis. The challenge, however, is how to plan for adaptation in a longer time horizon, i.e. 2030 and 2070. Based on the preliminary findings from land use models such as the LSA, APSIM and SGS pasture-growth models, the research team have geo-visualised two neighbouring farmscapes around Hamilton in the year 2030. These farmscapes have not been modelled on existing farms located within the region; rather, two hypothetical farmscapes have been created but both are based on existing production systems predominant in south-west Victoria, specifically wheat and sheep grazing.

Firstly a plan view map was produced using 3ds Max (3D modelling and animation package) and Photoshop (image editing package) of the two hypothetical farming systems. As shown in Figure 23.6 the farms are clearly divided into two parts by a central road to differentiate the faming practices in 2030. The plan view map of the hypothetical farms was overlayed in Google Earth to increase interactivity and ease of use for end-users. To make the geo-visualisation representation more realistic we prepared the 3D virtual environment of the above-described scenario within 3ds Max software packages. The farmscapes were created using a number of 3D models representing built infrastructure and vegetation (Pettit *et al.* 2009).

As shown in Figure 23.6, a number of panoramic image outputs were exported from 3ds Max representing multiple view points to cover maximum landscape coverage. These images where then imported in the Photo Overlay Creator software created by the University of College London's Centre for Advanced Spatial Analysis, to create 360° image spheres, and positioned at specific view points within Google Earth on top of the plan view maps of the farmscapes. These clickable spheres allowed users to explore the panoramic view from that point of the scenario. This interactive system helped end users to understand the complex science behind climate change in a very simple manner and improved the understanding towards adopting climate change technology in future farm practices.

According to Cullen *et al.* (2009), seasonal changes in the pasture growth pattern at Hamilton are characterised by a small increase in winter growth rates (caused by increased winter temperatures combined with elevated atmospheric carbon dioxide concentrations) and a decrease in the magnitude of both the spring peak growth rate and length of the spring growing season. The SGS pasture growth model indicated that spring is the season that is most affected because the rainfall reductions are projected to be largest at this time, as shown in Figure 23.7 (centre). The traditional farm with conventional farm practices faced low production in cropping and grazing (Figure 23.7 bottom). The loss of productivity has been attributed to higher temperatures, lesser rain with a subsequent further lowering of the water table, and more extreme events including drought and a greater number of frost days, the conditions for grazing have worsened with pasture productivity down and an associated reduction from 15 sheep per hectare to eight to 10 sheep per hectare.

The farm that has implemented climate change impact measures including planting deeper rooted perennials and hedgerows to reduce the impact of extreme weather conditions during

Figure 23.6: Plan view map (top) and image spheres of future farmscapes (bottom).

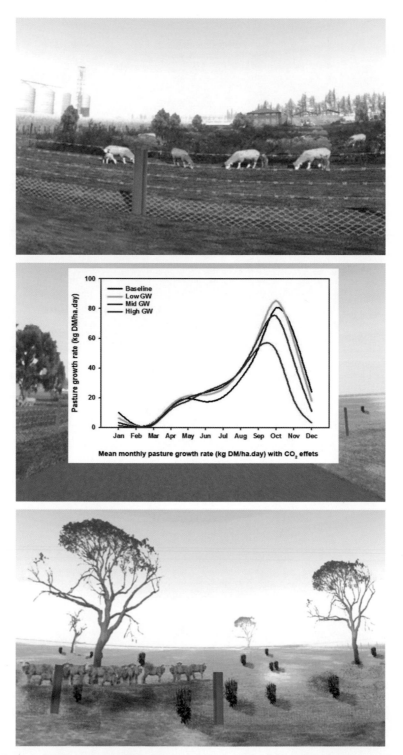

Figure 23.7: Geo-visualisation of future farming system under a changing climate as informed by the SGS model. Early technology adopter represented on the top, traditional farm on the bottom.

Figure 23.8: Renewable energy mix – solar, wind and biofuels.

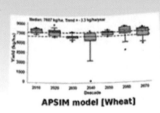

APSIM model [Wheat]

Figure 23.9: Additional tree planting to reduce water table, provide biodiversity habitat, and carbon credits.

lambing is illustrated in Figure 23.7 (top). This farm has also implemented a number of additional changes to the farming system including the establishment of a biofuel plant, wind turbines and solar panels as an alternative source of energy (see Figure 23.8). A biolink zone and bluegum plantation provide an important habitat to support biodiversity as well as carbon credits (see Figure 23.9).

Conclusions and future research

In this chapter we have presented some research in the field of geographical visualisation in communicating landscapes and farmscapes, both current and future as informed by climate change and land use change models and expert metrics. The next steps of the research will focus on the evaluation of the effectiveness of geographical visualisation as an approach for improving and bringing about societal and behavioural change. The research team will take the future farmscape geo-visualisation approach and apply it to the DemoDairy farm located in Terrang. Digital globe technology, 3D rendering software and the SIEVE computer gaming technology (Stock *et al.* 2008) will be applied to create current and future farmscapes for this operational dairy farm. The research team will also develop landscape geo-visualisation scenario storyline products for the south-west region. Initial feedback from end users of the geo-visualisation approach is quite promising and we believe such enhanced communications tools will assist in an accelerated understanding of climate change effects and adaptation strategies.

Acknowledgements

This research uses data provided by the Community Climate System Model project (www.ccsm.ucar.edu), supported by the Directorate for Geosciences of the National Science Foundation and the Office of Biological and Environmental Research of the US Department of Energy. Special thanks to Department of Primary Industries and Department of Sustainability and Environment for funding this research. Dr Claudia Pelizaro is thanked for her work on the LSA models and Angela Avery for her contributions in providing a farming system understanding used in refining the farmscape geo-visualisation products. Finally, we would like to thank the two anonymous reviewers for their time in reading an earlier version of the chapter and providing constructive feedback.

References

CSIRO and Bureau of Meteorology (2007) 'Climate change in Australia – technical report 2007'. CSIRO and Bureau of Meteorology, Australia. <www.climatechangeinaustralia.gov.au>

Cullen BR, Johnson IR, Eckard RJ, Lodge GM, Walker RG, Rawnsley RP and McCaskill MR (2009) Climate change impacts on Australian pasture systems. *Crop and Pasture Science* (**in press**).

Dockerty T, Lovett A, Sunnenberg G, Appleton K and Parry M (2005) Visualising the potential impacts of climate change on rural landscapes. *Computers, Environment and Urban Systems* **29**, 297–320.

Dykes J, MacEachren A and Kraak MJ (2005) *Exploring Geovisualization*. Elsevier, Oxford, UK.

FAO (1976) 'A Framework for Land Evaluation'. Soils Bulletin 22, Food and Agricultural Organisation (FAO), Rome.

Garnaut R (2008) *The Garnaut Climate Change Review*. Cambridge University Press, Melbourne.

IPCC (2007) Climate Change 2007: Synthesis Report. Contribution of Working Groups I, II and III to the Fourth Assessment Report of the Intergovernmental Panel on Climate Change. IPCC, Geneva, Switzerland.

Johnson IR, Lodge GM and White RE (2003) The Sustainable Grazing Systems Pasture Model: description, philosophy and application to the SGS National Experiment. *Australian Journal of Experimental Agriculture* **43**, 711–728, doi:10.1071/EA02213

Keating BA, Carberry PS, Hammer GL, Probert ME, Robertson MJ, Holzworth D, Huth NI, Hargreaves JNG, Meinke H, Hochman Z, McLean G, Verburg K, Snow V, Dimes JP, Silburn M, Wang E, Brown S, Bristow KL, Asseng S, Chapman S, McCown RL, Freebairn DM and Smith CJ (2003) An overview of APSIM, a model designed for farming systems simulation. *European Journal of Agronomy* **18**, 267–288.

Kelly PM and Adger WN (2000) Theory and practice in assessing vulnerability to climate change and facilitating adaptation. *Climate Change* **47**(4), 325–352.

Klosterman RE (1999) The what if? Collaborative planning support system. *Environment and Planning B: Planning and Design* **26**, 393–408.

Lovett A (2005) Futurescapes. *Computers, Environment and Urban Systems* **29**, 249–253.

Nourbakhsh I, Sargent R, Wright A, Cramer K, McClendon B and Jones M (2006) Mapping disaster zones. *Nature* **439** (7078), 787–788.

Pettit C, Cartwright W and Berry M (2006) Geographical visualization: a participatory planning support tool for imagining landscape futures. *Applied GIS* **2**(3), 22.1–22.17.

Pettit CJ, Sheth F, Harvey W and Cox M (2009) Building a 3D Object Library for Visualising Landscape Futures. In: *Proceedings of the 18th IMACS World Congress MODSIM09 Conference*, 13–17 July, Cairns, Queensland.

Saaty T (2000) *Fundamentals of Decision Making and Priority Theory with the Analytic Hierarchy Process*. RWS Publications, Pittsburgh.

Sheppard SRJ (2005) Landscape visualisation and climate change: the potential for influencing perceptions and behaviour. *Environmental Science and Policy* **8**, 637–654.

Skamarock WC, Klemp JB, Dudhia J, Gill DO, Barker DM, Wang W and Powers JG (2005) A Description of the Advanced Research WRF Version 2, NCAR Technical Note, NCAR/TN-468+STR.

Sposito V, Pelizaro C, Benke K, Anwar M, Rees D, Elsley M, O'Leary G, Wyatt R and Cullen B (2008) 'Climate Change Impacts on Agriculture and Forestry Systems in South West Victoria, Australia'. Department of Primary Industries, Melbourne.

Stock C, Bishop ID, O'Connor AN, Chen T, Pettit CJ and Aurambout J-P (2008) SIEVE: Collaborative Decision-making in an Immersive Online Environment. *Cartography and Geographic Information Science* **35**(2), 133–144(12).

Index

adaptation
  and agriculture 102–9
  climate change 23, 101, 102, 105–8, 109,
    110, 139, 155, 156, 163, 167–74, 186,
    193, 197, 198–202, 203, 206, 259–60,
    261–71
  drivers and benefits 198–9
  and freshwater biodiversity 79–81
  responses to infrastructure risks
    199–200
  and risk assessment 189–90, 200–2
  strategies and plans 209–17
adaptation planning 205–18
  evaluation criteria 209
  risk assessment 200–2
  role of evaluation 207–9
adaptive management 207–8
aerosols 31, 65–71
  and atmospheric circulation changes
    68–70
  data, models and experiments 68
  and ocean circulation changes 67–8
Agricultural Production Simulator
  (APSIM) 264–6
agriculture
  and climate change 249–56, 259–71
  deciles and climate variability 113,
    115–16
  future farmscapes 266–71
  growing season temperature 113, 115,
    118, 119, 121
  Western Australia 148–9
  see also cereal growing; wine grape
    growing
agriculture, climate change and
  Australian 101–10
  adaptation approaches 102–5
  consumption patterns 105–8
  and population growth 105–8
algal blooms 75
An Inconvenient Truth 249
arrowhead (Sagittaria montevidensis) 77

atmospheric circulation changes 65, 68–70,
  97
attitudes, individual 5–7, 10
Australian Climate Change Science
  Program (ACCSP) 31–2, 50, 97
  research highlights 36
  reviews 35
Australian landscapes and climate 74–5,
  249–56, 259–71

baroclinic instability changes 88–93
barriers to behaviour change 245–6
behavioural constraints and climate
  change 2–3
behavioural tendencies, innate and
  acquired 4–5
beliefs and values 5–7
biodiversity impacts 76–7
biofuels 182
Birchip Cropping Group (BCG) 251, 252,
  253
black swan (Cygnus atratus) 79
blackfish (Gadopsis marmoratus) 76
Brotherhood of St Laurence 155, 156
Bruntland Commission 149
Building and Sustainability Index
  (BASIX) 139, 140–1, 142
bulrush (Typha orientalis) 77
Bureau of Rural Sciences (BRS) 251, 252,
  254
bushfires 168

Canada 200
Cape Grim Baseline Air Pollution
  Station 31, 32
carbon dioxide, atmospheric
  concentrations 32, 208
Carbon Pollution Reduction Scheme
  (CPRS) 137, 141, 155, 189
  distributional impacts of 156–64
carp (Cyprinus carpio) 76
Centre for Low Emission Technology 237

cereal growing 115
China 20, 21
cities and climate change 177–83
climate and Australian landscapes 74–5,
    249–56, 259–71
climate change
    adaptation 23, 101, 102, 105–8, 109, 110,
        139, 155, 156, 163, 167–74, 186, 193,
        197, 198–202, 203, 206, 259–60,
        261–71
    and Australian agriculture 101–10,
        249–56, 259–71
    Australian landscapes and climate 74–5,
        249–56, 259–71
    challenges in talking about 224–6
    in our cities 177–83
    and climate knowledge 113–14
    communication tool 260–3
    formulation of attitudes 10
    and freshwater biodiversity 73–81
    future challenges 36–7
    future farmscapes 266–71
    future research 271
    and geo-visualisation 259–64, 266, 268,
        271
    holistic thinking 12
    and infrastructure 197–203
    models 261–6
    and multiple existing stressors 75–6
    newspaper coverage of 251
    public awareness and knowledge 237
    and resilience 13
    risk management 7–10, 185–6, 191–3
    scenarios and Google Earth 259–71
    sharing responsibility 10–11
    societal constraints 2–3, 242–6
    strategic thinking 12–13
    and sustainability 149–50
    trends 53, 250, 261–71
    and uncertainty 13
    what is acceptable? 7–11
    what is equitable? 11–12
    what is possible? 3–7
Climate Change in Australia 34, 119
climate change policy and the Great
    Crash 18, 19–22
climate change research highlights 32–6
    climate projections 34

communication 35–6
    modelling 33
    reviews 35
    sea level 34
    terrestrial changes 35
    tracking gas 32–3
    tracking the oceans 33–4
climate debate 145–7
climate influences, Australia 253, 254
climate modelling 32, 33, 34, 36, 37, 39–45
Climate Prediction Center Merged Analysis
    of Precipitation (CMAP) 40
climate projections 34
climate risk in human settlements 185–93
climate system, physical aspects of 8
climate variability 33, 34, 36, 101, 107, 113,
    114, 116, 133, 191, 197, 205, 207, 249, 250,
    252, 255
coastal ecosystems and freshwater
    biodiversity 77
communication and research 35–6
communication tools and climate
    change 260–3
Community Climate System Model 262,
    271
Copenhagen (2009) 17, 18, 25, 26, 45, 189
    and global recession 23–4
Coupled Global Climate Models
    (CGCMs) 39, 40
Coupled Model Intercomparison Project 39,
    40, 42, 43–9, 50, 86, 92, 97
CSIRO 31, 32–6, 61, 66, 67–70, 149, 193,
    206, 237–48, 249, 259

deciles and climate variability 113, 115–16
dryland farming and climate risk 113,
    114–15

economic policy 18, 21, 150
economics and energy technology 9
ecosystem processes and freshwater
    biodiversity 77–9
electric transit systems 178–81
electric hybrid vehicles 181–2
emissions growth and global recession
    19–21
emissions trading scheme (ETS) 18, 21,
    25–6, 141, 155, 156

energy
    efficiency 159–64
    renewable 21, 145, 150–1, 157, 177–8,
        181–2, 201, 241
    sustainable 145–51
energy efficient housing 137–42, 155–63
Energymark 237–48
    barriers to behaviour change 245–6
    key roles 239–40
    preliminary results 242–5
    process 239–40
    research measures 240–2
    session topics 240–6
evidence-based policy environment 208–9

farmscapes, climate change and future
    266–71
financial fragility 21
fish 76, 79
flooding 75, 168
fossil fuels 145, 147, 241
foxes 79
freshwater biodiversity and climate
    change 73–81
    adaptation and mitigation 79–81
    Australian landscapes and climate
        74–6
    biodiversity impacts 76–7
    ecosystem processes 77–9
    impacts on coastal ecosystems 77
    indirect impacts and interactions 79
freshwater ecosystems 73, 77, 81

Garnaut Climate Change Review 17–18, 19,
    24, 156, 157, 199, 224
geo-visualisation 259–64, 266, 268, 271
global annual mean surface temperature
    anomaly 146
Global Atmospheric Sampling Laboratory
    (GASLAB) 32
Global Carbon Project 45
global circulation models (GCMs) 47, 106,
    262, 263
global mean warming 32
global recession
    and the Australian policy
        discussion 24–6
    and Copenhagen 23–4

and emissions growth 19–21
    and investment in structural change 21
government energy efficiency programs
    163–4
Great Crash of 2008 17–18
    and climate change policy 19–22
Green Loans Program 163
GREENHOUSE 87 31, 35–6
greenhouse gases 2, 13, 17, 31, 32, 39, 53, 54,
    61, 65, 66, 70, 73, 86, 137, 140, 177, 182,
    198, 208, 241
growing season temperature and
    viticulture 113, 115, 118, 119, 121

heatwaves 123–33
Home Insulation Program (HIP) 163
households, low-income 156–7, 159–64
housing
    energy efficient 137–42, 155–63
    New South Wales 139
    sustainable 138–9
    Western Australia 137–42
human settlements and climate risk
    185–93

Indian Ocean 149
Indian Ocean Climate Initiative
    (IOCI) 53–7
infrastructure
    and climate change 197–203
    design standards 202
    materials selection 202
    risk assessment and adaptation
        planning 200–2
    risks, adaptation responses to 199–200
innovation 3, 115, 139, 142, 150–1, 177, 181,
    202, 226
    social 223–4, 233
    waves of 178
Integrated Assessment of Human
    Settlements (IAHS) 186, 189, 191
Intergovernmental Panel on Climate
    Change (IPCC) 3, 32, 60, 121, 145, 168,
    198, 208, 249, 259
invasive species 75, 76–7

knowledge transfer 3–4, 237
Kyoto protocol 17, 23, 25–6, 205, 241

land suitability analysis 171, 264
landscapes and climate, Australian 74–5,
    249–56, 259–71
lifestyles, zero-carbon 223–34
    and climate change 224–6
    lessons and conclusions 232–4
    reactions to 227–32
lippia (*Phyla canescens*) 77
liquefied natural gas (LNG) 182
Local Adaptation Pathways Program
    (LAPP) 186–7, 189, 191
low-income households 156–7, 159–64

Macquarie perch (*Macquaria*
    australasica) 76
Managing Climate Variability (MCV) 250
marine environments, rainfall patterns and
    interactions between freshwater and 77
market economy during and after
    recession 22
mean sea level pressure (MSLP) 40, 42, 43,
    50, 53, 54, 55, 58, 60, 61
monsoon 39–50
    Australian component 39
    changes in onset and duration 46–7
    model and validation datasets 40
    projected changes in tropical
        Australia 45–7
    seasonal variations 40–2, 50
monsoon, model-simulated climate 40–5
    interannual variability 44–5
    mean climate and circulation 43–4
    seasonal cycles over tropical
        Australia 40–3
mosquitofish (*Gambusia holbrooki*) 76–7
mountain shrimp (*Anaspides tasmaniae*) 76
Mt Lofty Ranges (SA) 167–74
Murray–Darling Basin 74, 114, 125–7, 130,
    167, 169

Nation Building Stimulus Plan 163
National Center for Atmospheric Research
    (NCAR) 57, 66, 69, 70
National Centers for Environmental
    Prediction (NCEP) 57, 86, 88, 89, 91,
    92–3, 97
National Climate Centre (NCC), Australian
    Bureau of Meteorology 40, 54, 58, 124

National Climate Change Adaptation
    Research Facility 80, 193, 206
National Coastal Vulnerability Assessment
    (NCVA) 186, 189
National Energy Efficiency Program
    (NEEP) 159, 163
National Tidal Centre 34
natural gas transport 182
natural resources management (NRM)
    167–74, 210
New South Wales
    and climate change 237
    housing 139, 163
    wine grape growing 125
New Zealand 200
newspaper coverage of climate change 251
Ningaloo Reef 149

ocean circulation changes 65, 66, 67–8, 71
ocean–atmosphere circulation 65–71
    and atmospheric circulation changes
        68–70
    changes 67–8
    data, models and experiments 68
oil vulnerability 177–83

peak oil 231
pedestrian-oriented development
    (POD) 178, 181
policy, global recession and Australian 24–6
population growth 185
    and agriculture 105–8
public awareness and knowledge of climate
    change 237
public transport 178–82, 230

Queensland 35, 50, 237

rail transport 178–80, 182
rainbow fish (*Melanotaenia* spp.) 76
rainbow trout (*Oncorhynchus mykiss*) 76
rainfall changes
    in northern Australia 75
    observed recent 54–7
    shifts in the synoptic systems 57–60
    in South Australia 116
rainfall decline 53, 54, 58, 60–1, 118, 253,
    256

rainfall patterns and interactions between freshwater and marine environments 77
rainfall trends in south-west Australia 53–61
  data 54
  observed recent rainfall changes 54–7
  rainfall decline 53, 54, 58, 60–1
  results 54–61
  shifts in the synoptic systems 57–60
regional vulnerability assessment 169
renewable energy 21, 145, 150–1, 157, 177, 178, 181–2, 201, 241
'Rising Above Hot Air' 224, 226, 234
risk assessment and adaptation 189–90, 191–2, 200–2
risk management 7–10, 185–6, 191–3, 200–2
road transportation 178–82

scepticism 4
sea level rise
  due to thermal expansion 32
  modelling and measurements 34
seasonal cycles over tropical Australia 40–3
seasonal variations (monsoon climates) 40–2, 50
silver perch (Bidyanus bidyanus) 76
societal constraints and climate change 2–3, 242–6
societal vulnerability 207
South Australia 113–21
  agriculture 251, 253–4, 261
  cereal growing 115
  and climate change 116, 251, 253–4, 261
  and climate change adaptation 167–74
  and climate information 116–21
  dryland farming 113, 114–15
  and extreme heat 123–33
  and household emissions 163
  Mt Lofty Ranges 167–74
  rainfall changes 116
  water management systems 167
  wine grape growing 113, 114–15, 123–4, 125, 133, 172
south-eastern Australia
  heatwave 123–33
southern hemisphere weather systems, changing 85–97
  baroclinic instability changes 88–93

circulation changes 86
  storm track modes 86–8
south-west Australia, rainfall trends in 53–61
spangled grunter (Leiotherapon unicolor) 76
Stern Review 249
storm track modes, changes in 86–8
storms 75, 85–6
sustainability and climate change 149–50
sustainable energy 145–51

Tasmania and wine grape growing 125, 130
terrestrial changes 35
transit-oriented development (TOD) 178, 180–1
transportation 178–83
  aviation 183
  and biofuels 182
  electric transit systems 178–82
  freight 182
  and natural gas 182
  and oil 177
  public 178–82, 230
  rail 178–80, 182
  regional 182
  road 178–82
tropical Australia
  change in mean climate 45
  change in seasonal climate 45–6
  interannual variability 45–6
  projected changes in 45–7
  seasonal cycles over 40–3
  see also monsoon; monsoon, model-simulated climate
tropical rainfall, change in seasonal cycle 45–6
trout cod (Maccullochella macquariensis) 76

unemployment 22
United Kingdom 79, 200–1, 249
United Nations Framework Convention on Climate Change (UNFCCC) 189, 205, 209
United States 200, 262

Victoria
  agriculture 251, 252–3, 253

and climate change 237, 239, 251, 252–3,
   253
housing 139, 163
south-west 259–71
wine grape growing 125–7, 129–30
and zero-carbon lifestyles 223–4
Victorian Centre for Climate Change
   Adaptation Research 193
Victorian Climate Change Adaptation
   Program (VCCAP) 259–60, 261
vineyards *see* wine grape growing

water efficiency 137–42
water hyacinth (*Eichhornia crassipes*) 77
water use in housing 137, 138
Watermark 239, 240
weather systems, changing southern
   hemisphere 85–97
   baroclinic instability changes 88–93
   circulation changes 86
   storm track modes 86–8
Western Australia
   agriculture 148–9, 253

climate change 147–9, 237, 249, 253
housing, energy and water
   efficiency 137–42
south-west (SWWA) 85, 86–7, 89, 93,
   147–8
wheat
   consumption 106
   production 105–7
   yields 103–4
wheat-based cropping 103–5
Wilkenfeld report 139, 140, 142
wine grape growing
   and extreme heat 123–33
   growing season temperature 113, 115,
      118, 119, 121
   New South Wales 125
   South Australia 113, 114–15, 123–4, 125,
      133, 172
   Victoria 125–7, 129–30
World Climate Research Program
   (WCRP) 39, 97

zero-carbon *see* lifestyles, zero-carbon